Digital Signal Processing

Digital Signal Processing

Wilson Newman

WILLFORD PRESS
www.willfordpress.com

Published by Willford Press,
118-35 Queens Blvd., Suite 400,
Forest Hills, NY 11375, USA

ISBN: 978-1-64728-331-5

Cataloging-in-Publication Data

Digital signal processing / Wilson Newman.
p. cm.
Includes bibliographical references and index.
ISBN 978-1-64728-331-5
1. Signal processing--Digital techniques. 2. Signal processing. 3. Digital communications.
4. Digital electronics. I. Newman, Wilson.
TK5102.9 .D54 2021
621.382 2--dc23

For information on all Willford Press publications
visit our website at www.willfordpress.com

Willford Press

Table of Contents

Preface

It is with great pleasure that I present this book. It has been carefully written after numerous discussions with my peers and other practitioners of the field. I would like to take this opportunity to thank my family and friends who have been extremely supporting at every step in my life.

The field of electrical engineering that focuses on modifying, analyzing and synthesizing signals such as images, sounds and biological measurements is referred to as signal processing. Digital signal processing refers to the utilization of digital processing for signal processing operations. The signals processed in this manner are an arrangement of numbers that represent the samples of a continuous variable in a domain such as space, time and frequency. There are a numerous fields where digital signal processing is applied such as biomedical engineering, telecommunications, seismology, control systems, digital image processing, spectral density estimation, speech processing, radar and sonar. It is useful in data compression as well as in error detection and correction. This book presents the complex subject of digital signal processing in the most comprehensible and easy to understand language. Most of the topics introduced herein cover new techniques and the applications of digital signal processing. This textbook is appropriate for students seeking detailed information in this area as well as for experts.

The chapters below are organized to facilitate a comprehensive understanding of the subject:

Chapter – Digital Signals and Systems

A signal which can be used for representation of data as a sequence of discrete values is known as a digital signal. Some of the important digital systems are linear systems, casual systems and linear time-invariant systems. This is an introductory chapter which will briefly introduce all the significant aspects of digital signals and systems.

Chapter – Understanding Digital Signal Processing

Digital signal processing deals with the diverse techniques which are used to improve the reliability and accuracy of digital connections. It uses computers and digital signal processors for processing the signals digitally. This chapter has been carefully written to provide an easy understanding of digital signal processing.

Chapter – Signal Sampling

The process of mapping input values from a large set to output values in a smaller set is referred to as signal sampling quantization. Analog-to-digital conversion and digital to analog conversion are the two main types of digital conversions. The topics elaborated in this chapter will help in gaining a better perspective about these types and related aspects of signal sampling quantization.

Chapter – Discrete Fourier Transform

Discrete Fourier transform refers to a method which is used to convert a finite sequence of functions which are equally-spaced into a same-length sequence of equally-spaced samples of the discrete-time Fourier transform. This chapter discusses discrete Fourier transform and signal spectrum in detail.

Chapter - Z-Transform

The Z-Transform is used to convert a discrete-time signal into a complex frequency-domain representation. Some of its important properties are linearity, time shifting, time expansion, convolution and time accumulation. This chapter closely examines these properties and related aspects of Z-Transform to provide an extensive understanding of the subject.

Chapter - Digital Filter and its Types

The development of a signal flow diagram which describes the filter in terms of operations on sample sequences is termed as digital filter realization. A few of the different types of filters are low-pass filters, high-pass filters, all-pass filters, anti-aliasing filters, etc. All these types of digital filter realizations and filters have been carefully analyzed in this chapter.

Wilson Newman

1

Digital Signals and Systems

A signal which can be used for representation of data as a sequence of discrete values is known as a digital signal. Some of the important digital systems are linear systems, casual systems and linear time-invariant systems. This is an introductory chapter which will briefly introduce all the significant aspects of digital signals and systems.

Digital Signals

Computers are digital in nature. Computers process, store, and communicate information in binary form, i.e. in the combination of 1s and 0s which has specific meaning in computer language. A binary digit (bit) is an individual 1 or O. Multiple bit streams are used in a computer network.

Contemporary computer systems communicate in binary mode through variations in electrical voltage. Digital signaling, in an electrical network, ‘involves a signal which varies in voltage to represent one of two discrete and well-defined states as depicted in figure such as either a positive (+) voltage and a null voltage (unipolar) or a positive (+) or a negative (-) voltage (bipolar).

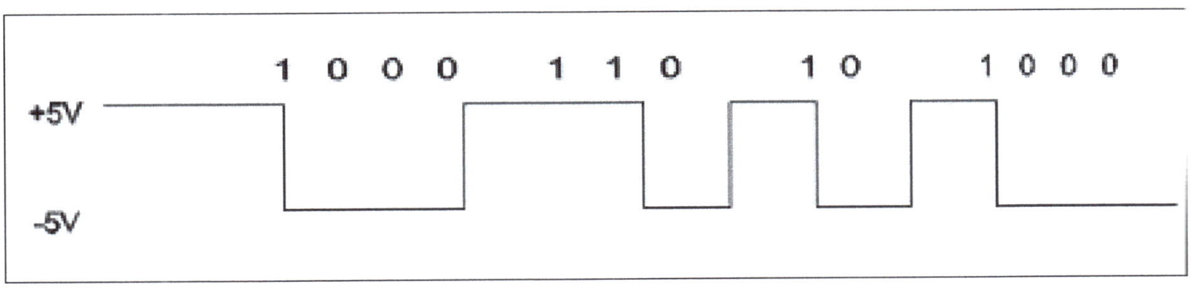

Although analog voice and video can be converted into digital, and digital data can be converted to analog, each format has its own advantages.

- It can have only a limited number of defined values such as 1 and 0.
- The transition of a digital signal from one value to other value is instantaneous.
- Digital signals are represented by square wave.

- In digital signals 1 is represented by having a positive voltage and 0 is represented by having no voltage or zero voltage as shown in figure.
- All the signals generated by computers and other digital devices are digital in nature.

Characteristics of Digital Signals

Bit Interval

It is the time required to send one single bit.

Bit rate

- It refers to the number of bit intervals in one second.
- Therefore bit rate is the number of bits sent in one second.
- Bit rate is expressed in bits per second (bps).
- Other units used to express bit rate are Kbps, Mbps and Gbps.
    - 1 kilobit per second (Kbps) = 1,000 bits per second.
    - 1 Megabit per second (Mbps) = 1,000,000 bits per second.
    - 1 Gigabit per second (Gbps) = 1,000,000,000 bits per second.

Advantages of Digital Signals

- Digital Data: Digital transmission certainly has the advantage where binary computer data is being transmitted. The equipment required to convert digital data to analog format and transmitting the digital bit streams over an analog network can be expensive, susceptible to failure, and can create errors in the information.
- Compression: Digital data can be compressed relatively easily, thereby increasing the efficiency of transmission. As a result, substantial volumes of voice, data, video and image information can be transmitted using relatively little raw bandwidth.
- Security: Digital systems offer better security. While analog systems offer some measure of security through the scrambling of several frequencies. Scrambling is fairly simple to defeat. Digital information, on the other hand, can be encrypted to create the appearance of a single, pseudorandom bit stream. Thereby, the true meaning of individual bits, sets of bits, or the total bit stream cannot be determined without having the key to unlock the encryption algorithm employed.
- Quality: Digital transmission offers improved error performance (quality) as compared to analog. This is due to the devices that boost the signal at periodic intervals in the transmission system in order to overcome the effects of attenuation. Additionally, digital networks deal more effectively with noise, which always is present in transmission networks.
- Cost: The cost of the computer components required in digital conversion and transmis-

sion has dropped considerably, while the ruggedness and reliability of those components has increased over the years.

- Upgradeability: Since digital networks are comprised of computer (digital) components, they are relatively easy to upgrade. Such upgrading can increase bandwidth, reduces the incidence of error and enhance functional value. Some upgrading can be effected remotely over a network, eliminating the need to dispatch expensive technicians for that purpose.
- Management: Generally speaking, digital networks can be managed much more easily and effectively due to the fact that such networks consist of computerized components. Such components can sense their own level of performance, isolate and diagnose failures, initiate alarms, respond to queries, and respond to commands to correct any failure.

Linear Systems

A system is called linear if it has two mathematical properties: homogeneity and additivity. If you can show that a system has both properties, then you have proven that the system is linear. Likewise, if you can show that a system doesn't have one or both properties, you have proven that it isn't linear. A third property, shift invariance, is not a strict requirement for linearity, but it is a mandatory property for most DSP techniques. When you see the term linear system used in DSP, you should assume it includes shift invariance unless you have reason to believe otherwise. These three properties form the mathematics of how linear system theory is defined and used.

As illustrated in figure, homogeneity means that a change in the input signal's amplitude results in a corresponding change in the output signal's amplitude. In mathematical terms, if an input signal of $x[n]$ results in an output signal of $y[n]$, an input of $k\,x[n]$ results in an output of $k\,y[n]$, for any input signal and constant, k.

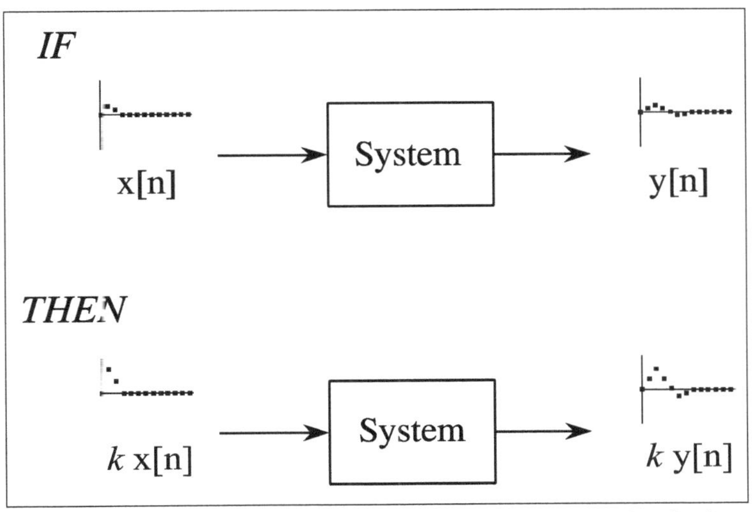

Definition of homogeneity. A system is said to be homogeneous if an amplitude change in the input results in an identical amplitude change in the output. That is, if x[n] results in y[n], then kx[n] results in ky[n], for any signal, x[n], and any constant, k.

A simple resistor provides a good example of both homogenous and nonhomogeneous systems. If the input to the system is the voltage across the resistor, $v(t)$, and the output from the system is the current through the resistor, $i(t)$, the system is homogeneous. Ohm's law guarantees this; if the voltage is increased or decreased, there will be a corresponding increase or decrease in the current. Now, consider another system where the input signal is the voltage across the resistor, $v(t)$, but the output signal is the power being dissipated in the resistor, $p(t)$. Since power is proportional to the square of the voltage, if the input signal is increased by a factor of two, the output signal is increase by a factor of four. This system is not homogeneous and therefore cannot be linear.

The property of additivity is illustrated in figure. Consider a system where an input of $y_1[n]$ produces an output of $y_1[n]$. Further suppose that a different input, $x_2[n]$, produces another output, $y_2[n]$. The system is said to be additive, if an input of $x_1[n]+x_2[n]$ results in an output of, $y_1[n]+y_2[n]$, for all possible input signals. In words, signals added at the input produce signals that are added at the output.

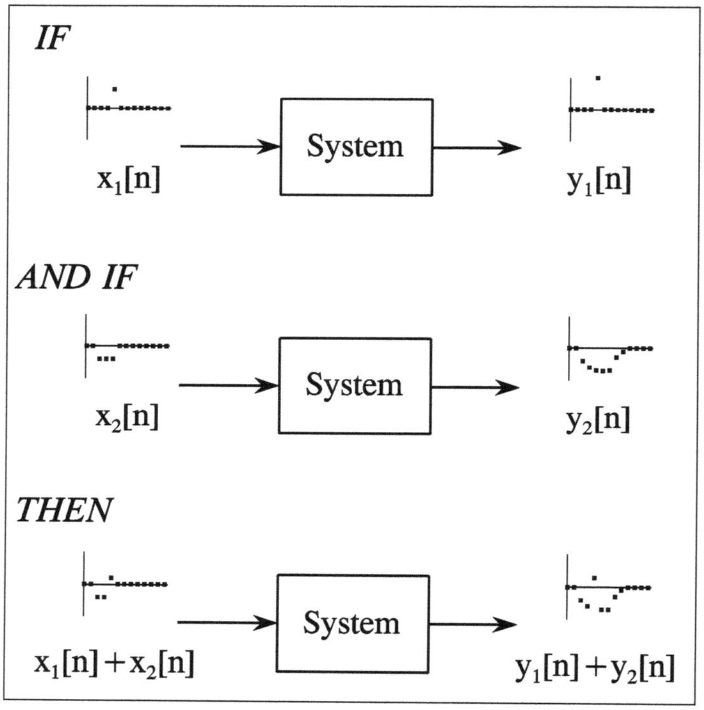

Definition of additivity. A system is said to be additive if added signals pass through it without interacting. Formally, if $x_1[n]$ results in $y_1[n]$, and if $x_2[n]$ results in $y_2[n]$, then $x_1[n]+x_2[n]$ results in $y_2[n]+y_2[n]$.

The important point is that added signals pass through the system without interacting. As an example, think about a telephone conversation with your Aunt Edna and Uncle Bernie. Aunt Edna begins a rather lengthy story about how well her radishes are doing this year. In the background, Uncle Bernie is yelling at the dog for having an accident in his favorite chair. The two voice signals are added and electronically transmitted through the telephone network. Since this system is additive, the sound you hear is the sum of the two voices as they would sound if transmitted individually.

A good example of a nonadditive circuit is the mixer stage in a radio transmitter. Two signals are present: an audio signal that contains the voice or music, and a carrier wave that can propagate through space when applied to an antenna. The two signals are added and applied to a nonlinearity, such as a pn junction diode. This results in the signals merging to form a third signal, a modulated radio wave capable of carrying the information over great distances.

As shown in figure, shift invariance means that a shift in the input signal will result in nothing more than an identical shift in the output signal. In more formal terms, if an input signal of $x[n]$ results in an output of $y[n]$, an input signal of $x[n+s]$ results in an output of $y[n+s]$, for any input signal and any constant, s. Pay particular notice to how the mathematics of this shift is written By adding a constant, s, to the independent variable, n, the waveform can be advanced or retarded in the horizontal direction. For example, when $s=2$, the signal is shifted left by two samples; when $s=-2$, the signal is shifted right by two samples.

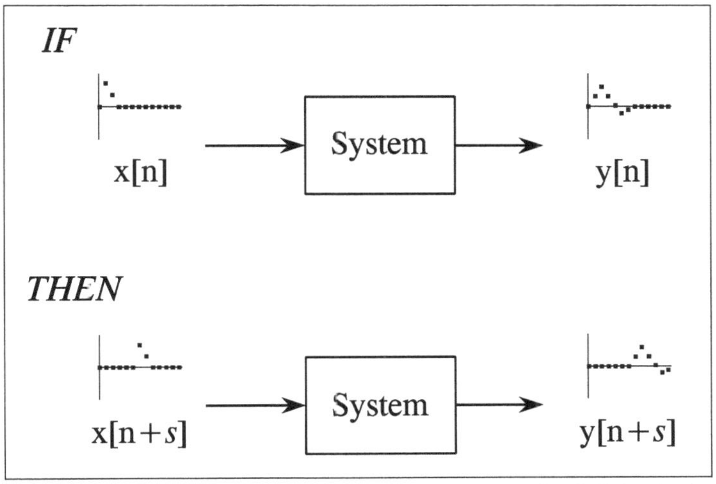

Definition of shift invariance. A system is said to be shift invariant if a shift in the input signal causes an identical shift in the output signal. In mathematical terms, if x[n] produces y[n], then x[n−s] produces y[n+s], for any signal, x[n], and any constant, s.

Shift invariance is important because it means the characteristics of the system do not change with time (or whatever the independent variable happens to be). If a blip in the input causes a blop in the output, you can be assured that another blip will cause an identical blop. Most of the systems you encounter will be shift invariant. This is fortunate, because it is difficult to deal with systems that change their characteristics while in operation. For example, imagine that you have designed a digital filter to compensate for the degrading effects of a telephone transmission line. Your filter makes the voices sound more natural and easier to understand. Much to your surprise, along comes winter and you find the characteristics of the telephone line have changed with temperature. Your compensation filter is now mismatched and doesn't work especially well. This situation may require a more sophisticated algorithm that can adapt to changing conditions.

Static Linearity and Sinusoidal Fidelity

Homogeneity, additivity, and shift invariance are important because they provide the mathematical basis for defining linear systems. Unfortunately, these properties alone don't provide most scientists and engineers with an intuitive feeling of what linear systems are about. The properties of static linearity and sinusoidal fidelity are often of help here. These are not especially important from a mathematical standpoint, but relate to how humans think about and understand linear systems.

Static linearity defines how a linear system reacts when the signals aren't changing, i.e., when they are DC or static. The static response of a linear system is very simple: the output is the input multiplied by a constant. That is, a graph of the possible input values plotted against the corresponding output values is a straight line that passes through the origin. This is shown in figure for two common linear systems: Ohm's law for resistors, and Hooke's law for springs. For comparison, figure shows the static relationship for two nonlinear systems: a pn junction diode, and the magnetic properties of iron.

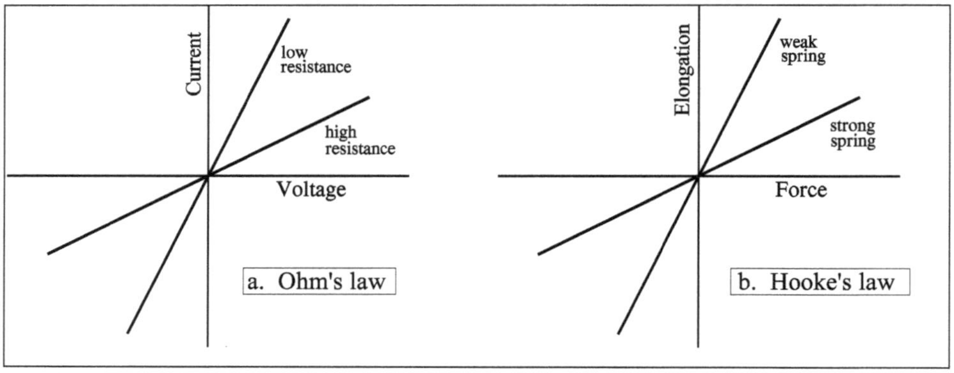

Two examples of static linearity. In (a), Ohm's law: the current through a resistor is equal to the voltage across the resistor divided by the resistance. In (b), Hooke's law: The elongation of a spring is equal to the applied force multiplied by the spring stiffness coefficient.

All linear systems have the property of static linearity. The opposite is usually true, but not always. There are systems that show static linearity, but are not linear with respect to changing signals. However, a very common class of systems can be completely understood with static linearity alone. In these systems it doesn't matter if the input signal is static or changing. These are called memoryless systems, because the output depends only on the present state of the input, and not on its history. For example, the instantaneous current in a resistor depends only on the instantaneous voltage across it, and not on how the signals came to be the value they are. If a system has static linearity, and is memoryless, then the system must be linear. This provides an important way to understand (and prove) the linearity of these simple systems.

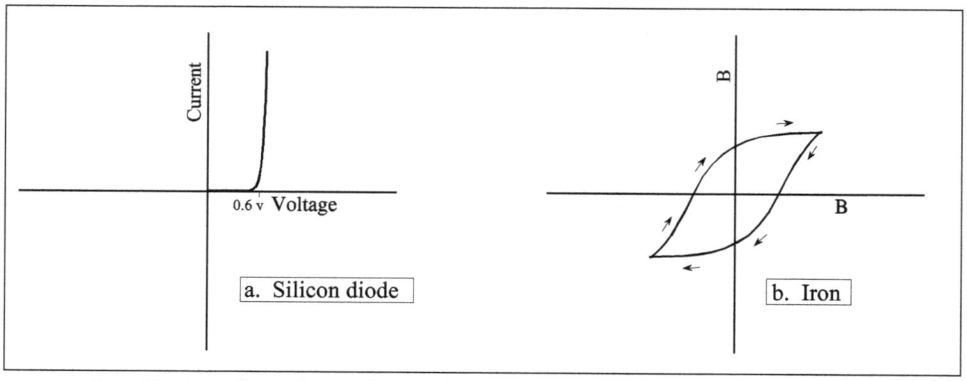

Two examples of DC nonlinearity. In (a), a silicon diode has an exponential relationship between voltage and current. In (b), the relationship between magnetic intensity, H, and flux density, B, in iron depends on the history of the sample, a behavior called hysteresis.

An important characteristic of linear systems is how they behave with sinusoids, a property we will call sinusoidal fidelity: If the input to a linear system is a sinusoidal wave, the output will also be a sinusoidal wave, and at exactly the same frequency as the input. Sinusoids are the only waveform that have this property. For instance, there is no reason to expect that a square wave entering a linear system will produce a square wave on the output. Although a sinusoid on the input guarantees a sinusoid on the output, the two may be different in amplitude and phase. This should be familiar from your knowledge of electronics: a circuit can be described by its frequency response, graphs of how the circuit's gain and phase vary with frequency.

Now for the reverse question: If a system always produces a sinusoidal output in response to a sinusoidal input, is the system guaranteed to be linear? The answer is no, but the exceptions are rare and usually obvious. For example, imagine an evil demon hiding inside a system, with the goal of trying to mislead you. The demon has an oscilloscope to observe the input signal, and a sine wave generator to produce an output signal. When you feed a sine wave into the input, the demon quickly measures the frequency and adjusts his signal generator to produce a corresponding output. Of course, this system is not linear, because it is not additive. To show this, place the sum of two sine waves into the system. The demon can only respond with a single sine wave for the output. This example is not as contrived as you might think; phase lock loops operate in much this way.

To get a better feeling for linearity, think about a technician trying to determine if an electronic device is linear. The technician would attach a sine wave generator to the input of the device, and an oscilloscope to the output. With a sine wave input, the technician would look to see if the output is also a sine wave. For example, the output cannot be clipped on the top or bottom, the top half cannot look different from the bottom half, there must be no distortion where the signal crosses zero, etc. Next, the technician would vary the amplitude of the input and observe the effect on the output signal. If the system is linear, the amplitude of the output must track the amplitude of the input. Lastly, the technician would vary the input signal's frequency, and verify that the output signal's frequency changes accordingly. As the frequency is changed, there will likely be amplitude and phase changes seen in the output, but these are perfectly permissible in a linear system. At some frequencies, the output may even be zero, that is, a sinusoid with zero amplitude. If the technician sees all these things, he will conclude that the system is linear. While this conclusion is not a rigorous mathematical proof, the level of confidence is justifiably high.

Examples of Linear and Nonlinear Systems

Provides examples of common linear and nonlinear systems. As you go through the lists, keep in mind the mathematician's view of linearity (homogeneity, additivity, and shift invariance), as well as the informal way most scientists and engineers use (static linearity and sinusoidal fidelity).

Examples of Linear Systems

- Wave propagation such as sound and electromagnetic waves.
- Electrical circuits composed of resistors, capacitors, and inductors.
- Electronic circuits, such as amplifiers and filters.
- Mechanical motion from the interaction of masses, springs, and dashpots (dampeners).

- Systems described by differential equations such as resistor-capacitor-inductor networks.
- Multiplication by a constant, that is, amplification or attenuation of the signal.
- Signal changes, such as echoes, resonances, and image blurring.
- The unity system where the output is always equal to the input.
- The null system where the output is always equal to the zero, regardless of the input.
- Differentiation and integration, and the analogous operations of first difference and running sum for discrete signals.
- Small perturbations in an otherwise nonlinear system, for instance, a small signal being amplified by a properly biased transistor.
- Convolution, a mathematical operation where each value in the output is expressed as the sum of values in the input multiplied by a set of weighing coefficients.
- Recursion, a technique similar to convolution, except previously calculated values in the output are used in addition to values from the input.

Examples of Nonlinear Systems

- Systems that do not have static linearity, for instance, the voltage and power in a resistor: $P = V^2R$, the radiant energy emission of a hot object depending on its temperature: $R = kT^4$, the intensity of light transmitted through a thickness of translucent material: $I = e^{-\alpha T}$, etc.
- Systems that do not have sinusoidal fidelity, such as electronics circuits for: peak detection, squaring, sine wave to square wave conversion, frequency doubling, etc.
- Common electronic distortion, such as clipping, crossover distortion and slewing.
- Multiplication of one signal by another signal, such as in amplitude modulation and automatic gain controls.
- Hysteresis phenomena, such as magnetic flux density versus magnetic intensity in iron, or mechanical stress versus strain in vulcanized rubber.
- Saturation, such as electronic amplifiers and transformers driven too hard.
- Systems with a threshold, for example, digital logic gates, or seismic vibrations that are strong enough to pulverize the intervening rock.
- Linear systems are defined by the properties of homogeneity, additivity, and shift invariance. Informally, most scientists and engineers think of linear systems in terms of static linearity and sinusoidal fidelity.

Special Properties of Linearity

Linearity is commutative, a property involving the combination of two or more systems. Figure shows the general idea. Imagine two systems combined in a cascade, that is, the output of one system is the input to the next. If each system is linear, then the overall combination will also be linear. The commutative property states that the order of the systems in the cascade can be rearranged without affecting the characteristics of the overall combination. You probably have used this principle in electronic circuits. For example, imagine a circuit composed of two stages, one for amplification, and one for filtering. Which is best, amplify and then filter, or filter and then amplify? If both stages are linear, the order doesn't make any difference and the overall result is the same. Keep in mind that actual electronics has nonlinear effects that may make the order important, for instance: interference, DC offsets, internal noise, slew rate distortion, etc.

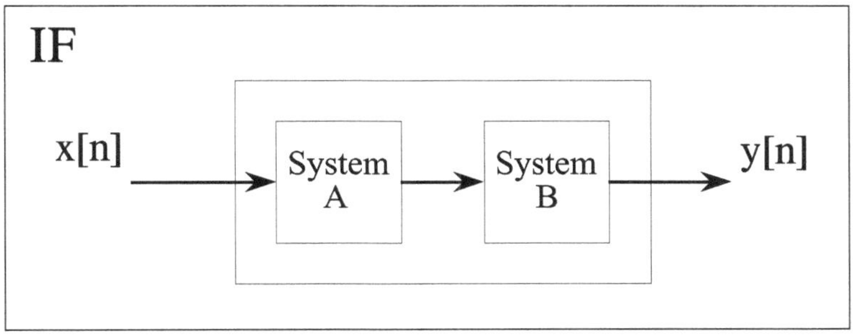

The commutative property for linear systems. When two or more linear systems are arranged in a cascade, the order of the systems does not affect the characteristics of the overall combination.

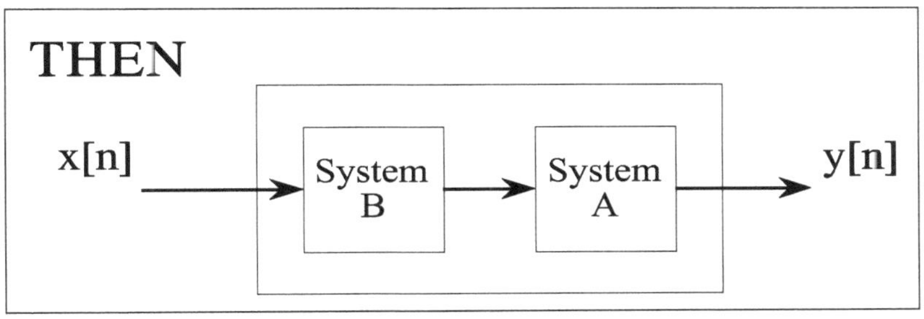

In figure shows above the next step in linear system theory: multiple inputs and outputs. A system with multiple inputs and outputs will be linear if it is composed of linear subsystems and additions of signals. The complexity does not matter, only that nothing nonlinear is allowed inside of the system.

To understand what linearity means for systems with multiple inputs and outputs, consider the following thought experiment. Start by placing a signal on one input while the other inputs are held at zero. This will cause the multiple outputs to respond with some pattern of signals. Next, repeat the procedure by placing another signal on a different input. Just as before, keep all of the other inputs at zero. This second input signal will result in another pattern of signals appearing on the multiple outputs. To finish the experiment, place both signals on their respective inputs simultaneously. The signals appearing on the outputs will simply be the superposition (sum) of the output signals produced when the input signals were applied separately.

The use of multiplication in linear systems is frequently misunderstood. This is because multiplication can be either linear or nonlinear, depending on what the signal is multiplied by. Figure illustrates the two cases. A system that multiplies the input signal by a constant, is linear. This system is an amplifier or an attenuator, depending if the constant is greater or less than one, respectively.

In contrast, multiplying a signal by another signal is nonlinear. Imagine a sinusoid multiplied by another sinusoid of a different frequency; the resulting waveform is clearly not sinusoidal.

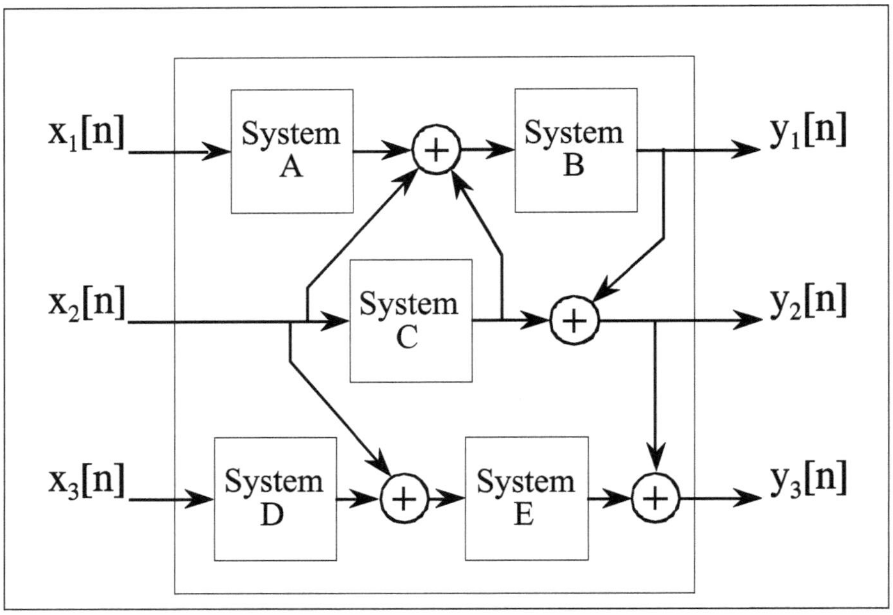

Any system with multiple inputs and outputs will be linear if it is composed of linear systems and signal additions.

Another commonly misunderstood situation relates to parasitic signals added in electronics, such as DC offsets and thermal noise. Is the addition of these extraneous signals linear or nonlinear? The answer depends on where the contaminating signals are viewed as originating. If they are viewed as coming from within the system, the process is nonlinear. This is because a sinusoidal input does not produce a pure sinusoidal output. Conversely, the extraneous signal can be viewed as externally entering the system on a separate input of a multiple input system. This makes the process linear, since only a signal addition is required within the system.

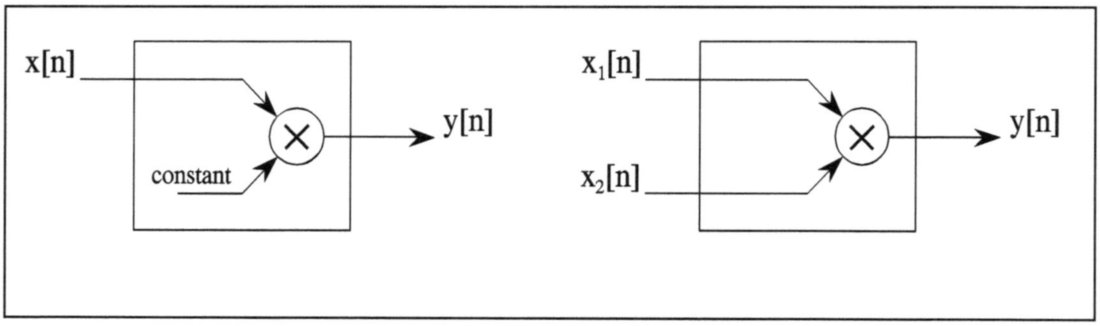

Linearity of multiplication. Multiplying a signal by a constant is a linear operation. In contrast, the multiplication of two signals is nonlinear.

Superposition: The Foundation of DSP

When we are dealing with linear systems, the only way signals can be combined is by scaling (multiplication of the signals by constants) followed by addition. For instance, a signal cannot be multiplied by another signal. Figure shows an example: three signals: $x_0[n]$, $x_1[n]$, *and* $x_2[n]$ are added to form a fourth signal, $x[n]$. This process of combining signals through scaling and addition is called synthesis.

Decomposition is the inverse operation of synthesis, where a single signal is broken into two or more additive components. This is more involved than synthesis, because there are infinite possible decompositions for any given signal. For example, the numbers 15 and 25 can only be synthesized (added) into the number 40. In comparison, the number 40 can be decomposed into $1+39$ or $2+38$ or $-30.5+60+10.5$, etc.

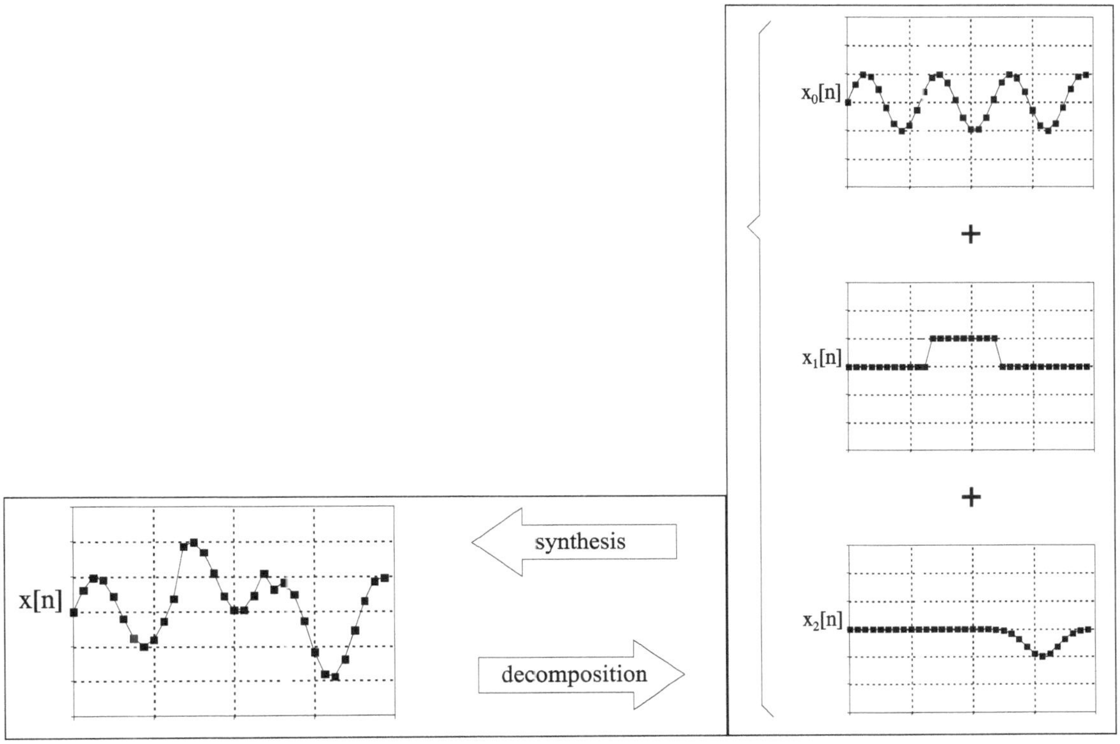

Illustration of synthesis and decomposition of signals. In synthesis, two or more signals are added to form another signal. Decomposition is the opposite process, breaking one signal into two or more additive component signals.

Now we come to the heart of DSP: superposition, the overall strategy for understanding how signals and systems can be analyzed. Consider an input signal, called $x[n]$, passing through a linear system, resulting in an output signal, $y[n]$. As illustrated in figure, the input signal can be decomposed into a group of simpler signals: $x_0[n]$ $x_1[n]$ $x_2[n]$ etc. We will call these the input signal components. Next, each input signal component is individually passed through the system, resulting in a set of output signal components: $y_0[n]$ $y_1[n]$ $y_2[n]$ etc. These output signal components are then synthesized into the output signal, y[n].

Here is the important part: the output signal obtained by this method is identical to the one produced by directly passing the input signal through the system. This is a very powerful idea. Instead of trying to understanding how complicated signals are changed by a system, all we need to know is how simple signals are modified. In the jargon of signal processing, the input and output signals are viewed as a superposition (sum) of simpler waveforms. This is the basis of nearly all signal processing techniques.

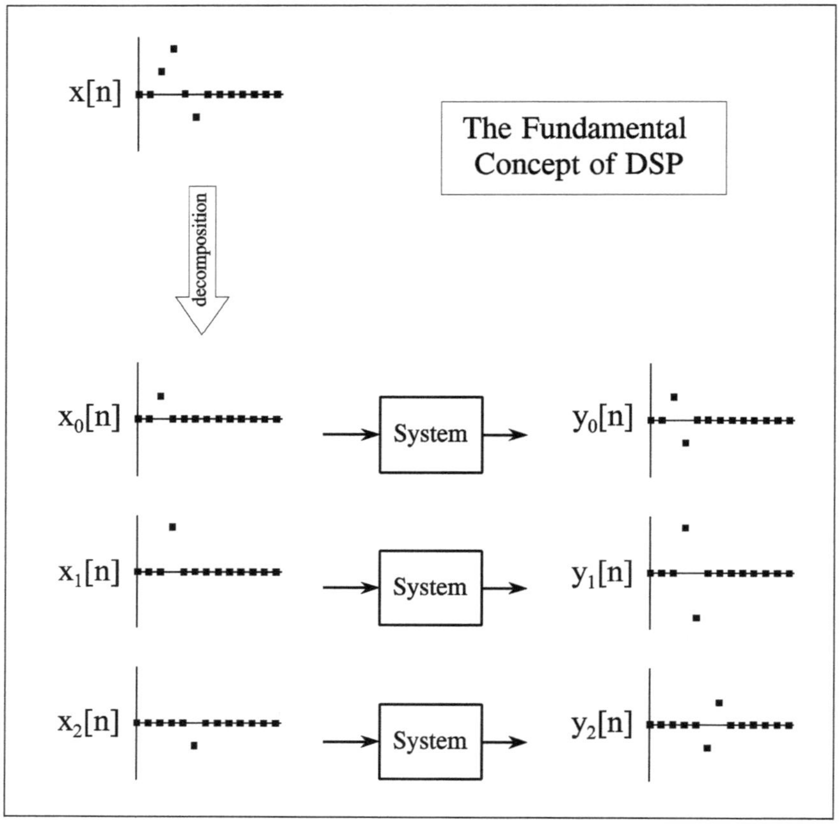

The fundamental concept in DSP. Any signal, such as $x[n]$, can be decomposed into a group of additive components, shown here by the signals: $x_1[n]$, $x_2[n]$, and $x_3[n]$ these components through a linear system produces the signals, $y_1[n]$, and $y_2[n]$, $y_3[n]$ The synthesis (addition) of these output signals forms $y_3[n]$, the same signal produced when x [n] is passed through the system.

As a simple example of how superposition is used, multiply the number 2041 by the number 4, in your head. How did you do it? You might have imagined 2041 match sticks floating in your mind, quadrupled the mental image, and started counting. Much more likely, you used superposition to simplify the problem. The number 2041 can be decomposed into: 2000 + 40 + 1 . Each of these components can be multiplied by 4 and then synthesized to find the final answer, i.e., 8000 + 160 + 4 = 8164 .

Common Decompositions

Keep in mind that the goal of this method is to replace a complicated problem with several easy ones. If the decomposition doesn't simplify the situation in some way, then nothing has been gained. There are two main ways to decompose signals in signal processing: impulse decomposition and Fourier decomposition. In addition, several minor decompositions are occasionally used. Here are brief descriptions of the two major decompositions, along with three of the minor ones.

Impulse Decomposition

As shown in figure, impulse decomposition breaks an N samples signal into N component signals, each containing N samples. Each of the component signals contains one point from the original signal, with the remainder of the values being zero. A single nonzero point in a string of zeros is called an impulse. Impulse decomposition is important because it allows signals to be examined one sample at a time. Similarly, systems are characterized by how they respond to impulses. By knowing how a system responds to an impulse, the system's output can be calculated for any given input.

Step Decomposition

Step decomposition, shown in figure, also breaks an N sample signal into N component signals, each composed of N samples. Each component signal is a step, that is, the first samples have a value of zero, while the last samples are some constant value. Consider the decomposition of an N point signal, x[n], into the components: $x_0[n]$, $x_1[n]$, $x_2[n], \cdots x_{N-1}[n]$ The kth component signal, $x_k[n]$ is composed of zeros for points 0 through $k-1$, while the remaining points have a value of: $x[k]$ $x[k-1]$ For example, the 5th component signal, $x_5[n]$, is composed of zeros for points 0 through 4, while the remaining samples have a value of: $x[5]$ $x[4]$ (the difference between sample 4 and 5 of the original signal). As a special case, $x_0[n]$ has all of its 0 [n] samples equal to $x[0]$. Just as impulse decomposition looks at signals one point at a time, step decomposition characterizes signals by the difference between adjacent samples. Likewise, systems are characterized by how they respond to a change in the input signal.

Even/Odd Decomposition

The even/odd decomposition, shown in figure, breaks a signal into two component signals, one having even symmetry and the other having odd symmetry. An N point signal is said to have even symmetry if it is a mirror image around point $N/2$. That is, sample $x[N/2+1]$ must equal $x[N/2-1]$ sample $x[N/2+2]$, etc. Similarly, odd symmetry occurs when the matching points have equal magnitudes but are opposite in sign, such as: $x[N/2+1] = -x[N/2-1]$, $x[N/2+2] = -x[N/2-2]$, etc. These definitions assume that the signal is composed of an even number of samples, and that the indexes run from 0 to $N-1$. The decomposition is calculated from the relations:

$$x_E[n] = \frac{x[n]+x[N-n]}{2}$$

$$x_o[n] = \frac{x[n]-x[N-n]}{2}$$

This may seem a strange definition of left-right symmetry, since $N/2$ ½ (between two samples) is the exact center of the signal, not $N/2$. Likewise, this off-center symmetry means that sample zero needs special handling. What's this all about?

This decomposition is part of an important concept in DSP called circular symmetry. It is based on viewing the end of the signal as connected to the beginning of the signal. Just as point $x[4]$ is next to point $x[5]$, point $x[N-1]$ is next to point $x[0]$. Picture a snake biting its own tail. When even and odd signals are viewed in this circular manner, there are actually two lines of symmetry, one at point $x[N/2]$ and another at point $x[0]$. For example, in an even signal, this symmetry around $x[0]$ means that point $x[1]$ equals point $x[N-1]$, point $x[2]$ equals point $x[N-2]$, etc. In an odd signal, point 0 and point $N/2$ always have a value of zero. In an even signal, point 0 and point $N/2$ are equal to the corresponding points in the original signal.

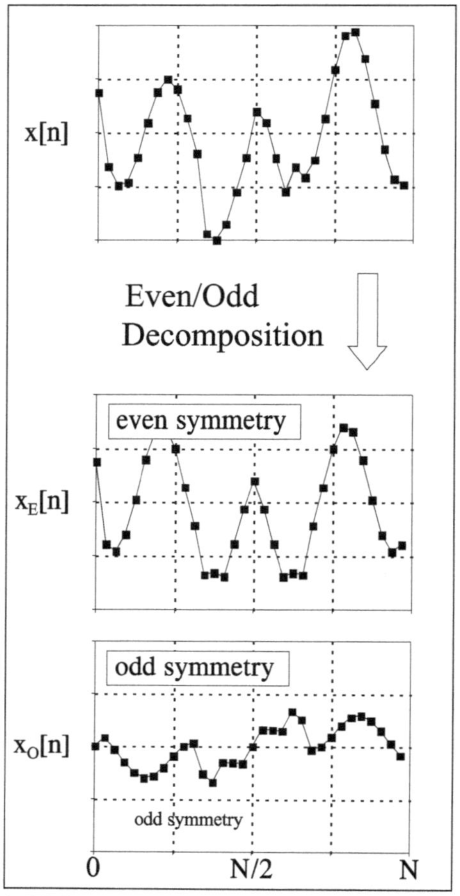

Example of even/odd decomposition. An N point signal is broken into two N point signals, one with even symmetry, and the other with odd symmetry.

What is the motivation for viewing the last sample in a signal as being next to the first sample? There is nothing in conventional data acquisition to support this circular notion. In fact, the first and last samples generally have less in common than any other two points in the sequence. It's common sense. The missing piece to this puzzle is a DSP technique called Fourier analysis. The mathematics of Fourier analysis inherently views the signal as being circular, although it usually has no physical meaning in terms of where the data came from. For now, the important thing to understand is that Eq. 5-1 provides a valid decomposition, simply because the even and odd parts can be added together to reconstruct the original signal.

Interlaced Decomposition, the interlaced decomposition breaks the signal into two component signals, the even sample signal and the odd sample signal. To find the even sample signal, start with the original signal and set all of the odd numbered samples to zero. To find the odd sample signal, start with the original signal and set all of the even numbered samples to zero. It's that simple.

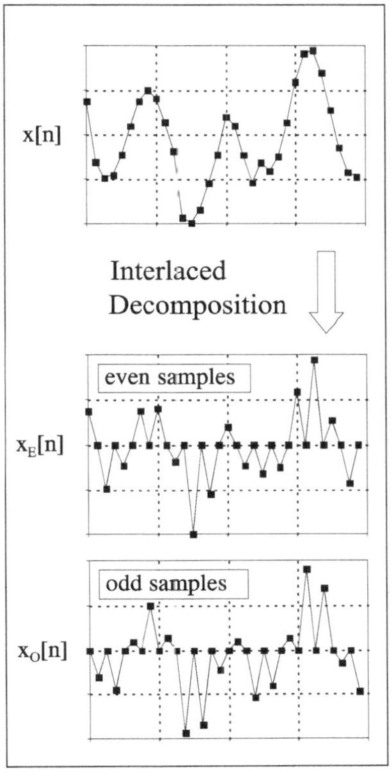

Example of interlaced decomposition. An N point signal is broken into two N point signals, one with the odd samples set to zero, the other with the even samples set to zero.

At first glance, this decomposition might seem trivial and uninteresting. This is ironic, because the interlaced decomposition is the basis for an extremely important algorithm in DSP, the Fast Fourier Transform (FFT). The procedure for calculating the Fourier decomposition has been known for several hundred years. Unfortunately, it is frustratingly slow, often requiring minutes or hours to execute on present day computers. The FFT is a family of algorithms developed in the 1960s to reduce this computation time. The strategy is an exquisite example of DSP: reduce the signal to elementary components by repeated use of the interlace transform; calculate the Fourier decomposition of the individual components; synthesize the results into the final answer. The results are dramatic; it is common for the speed to be improved by a factor of hundreds or thousands.

Fourier Decomposition Fourier decomposition is very mathematical and not at all obvious. Figure shows an example of the technique. Any N point signal can be decomposed into $N+2$ signals, half of them sine waves and half of them cosine waves. The lowest frequency cosine wave (called $x_{C0}[n]$ in this illustration), makes zero complete cycles over the N samples, i.e., it is a DC signal. The next cosine components: $x_{C1}[n], x_{C2}[n]$, and $x_{C3}[n]$, make 1, 2, and 3 complete cycles over the N samples, respectively. This pattern holds for the remainder of the cosine waves, as well as for the sine wave components. Since the frequency of each component is fixed, the only thing that changes for different signals being decomposed is the amplitude of each of the sine and cosine waves.

Fourier decomposition is important for three reasons. First, a wide variety of signals are inherently created from superimposed sinusoids. Audio signals are a good example of this. Fourier decomposition provides a direct analysis of the information contained in these types of signals. Second, linear systems respond to sinusoids in a unique way: a sinusoidal input always results in a sinusoidal output. In this approach, systems are characterized by how they change the amplitude and phase of sinusoids passing through them. Since an input signal can be decomposed into sinusoids, knowing how a system will react to sinusoids allows the output of the system to be found. Third, the Fourier decomposition is the basis for a broad and powerful area of mathematics called Fourier analysis, and the even more advanced Laplace and z-transforms. Most cutting-edge DSP algorithms are based on some aspect of these techniques.

Why is it even possible to decompose an arbitrary signal into sine and cosine waves? How are the amplitudes of these sinusoids determined for a particular signal? What kinds of systems can be designed with this technique? The details of the Fourier decomposition are too involved to be presented in this brief overview. For now, the important idea to understand is that when all of the component sinusoids are added together, the original signal is exactly reconstructed.

Alternatives to Linearity

To appreciate the importance of linear systems, consider that there is only one major strategy for analyzing systems that are nonlinear. That strategy is to make the nonlinear system resemble a linear system. There are three common ways of doing this:

First, ignore the nonlinearity. If the nonlinearity is small enough, the system can be approximated as linear. Errors resulting from the original assumption are tolerated as noise or simply ignored.

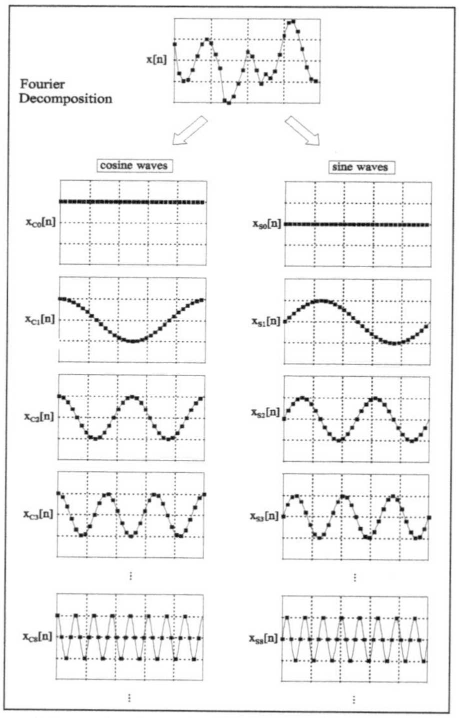

Illustration of Fourier decomposition. An N point signal is decomposed into N+2 signals, each having N points. Half of these signals are cosine waves, and half are sine waves. The frequencies of the sinusoids are fixed; only the amplitudes can change.

Second, keep the signals very small. Many nonlinear systems appear linear if the signals have a very small amplitude. For instance, transistors are very nonlinear over their full range of operation, but provide accurate linear amplification when the signals are kept under a few millivolts. Operational amplifiers take this idea to the extreme. By using very high open-loop gain together with negative feedback, the input signal to the op amp (i.e., the difference between the inverting and noninverting inputs) is kept to only a few microvolts. This minuscule input signal results in excellent linearity from an otherwise ghastly nonlinear circuit.

Third, apply a linearizing transform. For example, consider two signals being multiplied to make a third: $a[n] = b[n] \times c[n]$. Taking the logarithm of the signals changes the nonlinear process of multiplication into the linear process of addition: $\log(a[n]) = \log(b[n]) + \log(c[n])$. The fancy name for this approach is homomorphic signal processing. For example, a visual image can be modeled as the reflectivity of the scene (a two-dimensional signal) being multiplied by the ambient illumination (another two-dimensional signal). Homomorphic techniques enable the illumination signal to be made more uniform, thereby improving the image.

the two main techniques of signal processing: convolution and Fourier analysis. Both are based on the strategy presented in this chapter: (1) decompose signals into simple additive components, (2) process the components in some useful manner, and (3) synthesize the components into a final result. This is DSP.

Linear Time-invariant System

Linear time-invariant theory, commonly known as LTI system theory, investigates the response of a linear and time-invariant system to an arbitrary input signal. Trajectories of these systems are commonly measured and tracked as they move through time (e.g., an acoustic waveform), but in applications like image processing and field theory, the LTI systems also have trajectories in spatial dimensions. Thus, these systems are also called *linear translation-invariant* to give the theory the most general reach. In the case of generic discrete-time (i.e., sampled) systems, *linear shift-invariant* is the corresponding term. A good example of LTI systems are electrical circuits that can be made up of resistors, capacitors, and inductors.. It has been used in applied mathematics and has direct applications in NMR spectroscopy, seismology, circuits, signal processing, control theory, and other technical areas.

The defining properties of any LTI system are *linearity* and *time invariance*.

- *Linearity* means that the relationship between the input and the output of the system is a linear map: If input $x_1(t)$ produces response $y_1(t)$, and input $x_2(t)$ produces response $y_2(t)$, then the *scaled* and *summed* input $a_1x_1(t) + a_2x_2(t)$ produces the scaled and summed response $a_1y_1(t) + a_2y_2(t)$ where a_1 and a_2 are real scalars. It follows that this can be extended to an arbitrary number of terms, and so for real numbers $c_1, c_2, \ldots, c_k$, ,

  Input $\sum_k c_k x_k(t)$ produces output $\sum_k c_k y_k(t)$.

In particular,

Input $\int_{-\infty}^{\infty} c_\omega x_\omega(t) \mathrm{d}\omega$ produces output $\int_{-\infty}^{\infty} c_\omega y_\omega(t) \mathrm{d}\omega$

where c_ω and x_ω are scalars and inputs that vary over a continuum indexed by ω. Thus if an input function can be represented by a continuum of input functions, combined "linearly", as shown, then the corresponding output function can be represented by the corresponding continuum of output functions, *scaled* and *summed* in the same way.

- Time invariance means that whether we apply an input to the system now or T seconds from now, the output will be identical except for a time delay of T seconds. That is, if the output due to input $x(t)$ is $y(t)$, then the output due to input $x(t-T)$ is $y(t-T)$. Hence, the system is time invariant because the output does not depend on the particular time the input is applied.

The fundamental result in LTI system theory is that any LTI system can be characterized entirely by a single function called the system's impulse response. The output of the system is simply the convolution of the input to the system with the system's impulse response. This method of analysis is often called the *time domain* point-of-view. The same result is true of discrete-time linear shift-invariant systems in which signals are discrete-time samples, and convolution is defined on sequences.

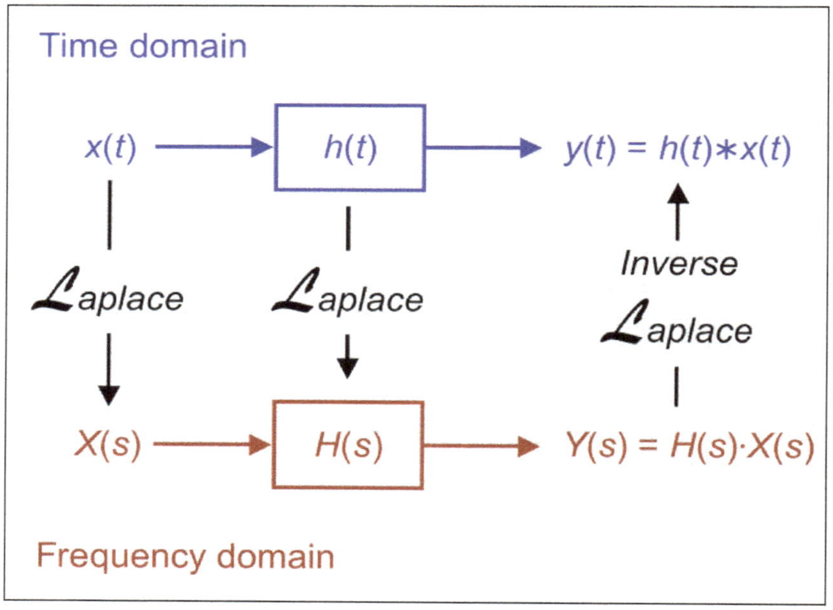

Relationship between the time domain and the frequency domain.

Equivalently, any LTI system can be characterized in the *frequency domain* by the system's transfer function, which is the Laplace transform of the system's impulse response (or Z transform in the case of discrete-time systems). As a result of the properties of these transforms, the output of the system in the frequency domain is the product of the transfer function and the transform of the input. In other words, convolution in the time domain is equivalent to multiplication in the frequency domain.

For all LTI systems, the eigenfunctions, and the basis functions of the transforms, are complex exponentials. This is, if the input to a system is the complex waveform $A_s e^{st}$ for some complex amplitude A_s and complex frequency s, the output will be some complex constant times the input, say $B_s e^{st}$ for some new complex amplitude B_s. The ratio B_s / A_s is the transfer function at frequency s.

Since sinusoids are a sum of complex exponentials with complex-conjugate frequencies, if the input to the system is a sinusoid, then the output of the system will also be a sinusoid, perhaps with a different amplitude and a different phase, but always with the same frequency upon reaching steady-state. LTI systems cannot produce frequency components that are not in the input.

LTI system theory is good at describing many important systems. Most LTI systems are considered "easy" to analyze, at least compared to the time-varying and nonlinear case. Any system that can be modeled as a linear homogeneous differential equation with constant coefficients is an LTI system. Examples of such systems are electrical circuits made up of resistors, inductors, and capacitors (RLC circuits). Ideal spring–mass–damper systems are also LTI systems, and are mathematically equivalent to RLC circuits.

Most LTI system concepts are similar between the continuous-time and discrete-time (linear shift-invariant) cases. In image processing, the time variable is replaced with two space variables, and the notion of time invariance is replaced by two-dimensional shift invariance. When analyzing filter banks and MIMO systems, it is often useful to consider vectors of signals.

A linear system that is not time-invariant can be solved using other approaches such as the Green function method. The same method must be used when the initial conditions of the problem are not null.

Continuous-time Systems

Impulse Response and Convolution

The behavior of a linear, continuous-time, time-invariant system with input signal $x(t)$ and output signal $y(t)$ is described by the convolution integral:

$$y(t) = (x * h)(t) \stackrel{\text{def}}{=} \int_{-\infty}^{\infty} x(t-\tau)\cdot h(\tau)\,\mathrm{d}\tau$$

$$= \int_{-\infty}^{\infty} x(\tau)\cdot h(t-\tau)\,\mathrm{d}\tau, \text{ (using commutativity)}$$

where $h(t)$ is the system's response to an impulse: $x(\tau) = \delta(\tau)$. $y(t)$ is therefore proportional to a weighted average of the input function $x(\tau)$. The weighting function is $h(-\tau)$, simply shifted by amount t. As t changes, the weighting function emphasizes different parts of the input function. When $h(\tau)$ is zero for all negative τ, $y(t)$ depends only on values of x prior to time t and the system is said to be causal.

To understand why the convolution produces the output of an LTI system, let the notation $\{x(u-\tau);\ u\}$ represent the function $x(u-\tau)$ with variable u and constant τ. And let the shorter notation $\{x\}$ represent $\{x(u);\ u\}$. Then a continuous-time system transforms an input function, $\{x,\}$

into an output function, $\{y\}$. And in general, every value of the output can depend on every value of the input. This concept is represented by:

$$y(t) \stackrel{\text{def}}{=} O_t\{x\},$$

where O_t is the transformation operator for time t. In a typical system, $y(t)$ depends most heavily on the values of x that occurred near time t Unless the transform itself changes with t the output function is just constant, and the system is uninteresting.

For a linear system, O must satisfy eq. Input $\int_{-\infty}^{\infty} c_\omega x_\omega(t)\mathrm{d}\omega$ produces output $\int_{-\infty}^{\infty} c_\omega y_\omega(t)\mathrm{d}\omega$:

$$O_t\left\{\int_{-\infty}^{\infty} c_\tau\, x_\tau(u)\mathrm{d}\tau;\ u\right\} = \int_{-\infty}^{\infty} c_\tau\, \underbrace{y_\tau(t)}_{O_t\{x_\tau\}}\mathrm{d}\tau.$$

And the time-invariance requirement is:

$$\begin{aligned} O_t\{x(u-\tau); u\} &= y(t-\tau) \\ &\stackrel{\text{def}}{=} O_{t-\tau}\{x\}. \end{aligned}$$

In this notation, we can write the impulse response as $h(t) \stackrel{\text{def}}{=} O_t\{\delta(u);\ u\}$.

Similarly:

$$\begin{aligned} h(t-\tau) &\stackrel{\text{def}}{=} O_{t-\tau}\{\delta(u);\ u\} \\ &= O_t\{\delta(u-\tau);\ u\}. \end{aligned}$$

Substituting this result into the convolution integral:

$$\begin{aligned} (x*h)(t) &= \int_{-\infty}^{\infty} x(\tau)\cdot h(t-\tau)\mathrm{d}\tau \\ &= \int_{-\infty}^{\infty} x(\tau)\cdot O_t\{\delta(u-\tau); u\}\mathrm{d}\tau, \end{aligned}$$

which has the form of the right side of eq. $O_t\left\{\int_{-\infty}^{\infty} c_\tau\, x_\tau(u)\mathrm{d}\tau;\ u\right\} = \int_{-\infty}^{\infty} c_\tau\, \underbrace{y_\tau(t)}_{O_t\{x_\tau\}}\mathrm{d}\tau.$ for the case $c_\tau = x(\tau)$ and $x_\tau(u) = \delta(u-\tau)$. Eq. then allows this continuation:

$$\begin{aligned} (x*h)(t) &= O_t\left\{\int_{-\infty}^{\infty} x(\tau)\cdot\delta(u-\tau)\mathrm{d}\tau; u\right\} \\ &= O_t\{x(u); u\} \\ &\stackrel{\text{def}}{=} y(t). \end{aligned}$$

In summary, the input function, $\{x\}$, can be represented by a continuum of time-shifted impulse functions, combined "linearly", as shown at eq. Input $\int_{\infty}^{\infty} c_\omega x_\omega(t)\mathrm{d}$ produces output $\int_{-\infty}^{\infty} c_\omega y_\omega(t)\mathrm{d}\omega$

The system's linearity property allows the system's response to be represented by the corresponding continuum of impulse responses, combined in the same way. And the time-invariance property allows that combination to be represented by the convolution integral.

The mathematical operations above have a simple graphical simulation.

Exponentials as Eigenfunctions

An eigenfunction is a function for which the output of the operator is a scaled version of the same function. That is,

$$\mathcal{H}f = \lambda f,$$

where f is the eigenfunction and λ is the eigenvalue, a constant.

The exponential functions Ae^{st}, where $A, s \in \mathbb{C}$, are eigenfunctions of a linear, time-invariant operator. A simple proof illustrates this concept. Suppose the input is $x(t) = Ae^{st}$. The output of the system with impulse response $h(t)$ is then,

$$\int_{-\infty}^{\infty} h(t-\tau) Ae^{s\tau}\, \mathrm{d}\tau$$

which, by the commutative property of convolution, is equivalent to,

$$\begin{aligned} \overbrace{\int_{-\infty}^{\infty} h(\tau) Ae^{s(t-\tau)}\, \mathrm{d}\tau}^{\mathcal{H}f} &= \int_{-\infty}^{\infty} h(\tau) Ae^{st} e^{-s\tau}\, \mathrm{d}\tau \\ &= Ae^{st} \int_{-\infty}^{\infty} h(\tau) e^{-s\tau}\, \mathrm{d}\tau \\ &= \underbrace{\overbrace{Ae^{st}}^{f}}_{\text{Input}} \underbrace{\overbrace{H(s)}^{\lambda}}_{\text{Scalar}}, \end{aligned}$$

where the scalar,

$$H(s) \stackrel{\text{def}}{=} \int_{-\infty}^{\infty} h(t) e^{-st}\, \mathrm{d}t$$

is dependent only on the parameter s.

So the system's response is a scaled version of the input. In particular, for any $A, s \in \mathbb{C}$, the system output is the product of the input Ae^{st} and the constant $H(s)$. Hence, Ae^{st} is an eigenfunction of an LTI system, and the corresponding eigenvalue is $H(s)$.

Direct Proof

It is also possible to directly derive complex exponentials as eigenfunctions of LTI systems.

Let's set $v(t) = e^{i\omega t}$ some complex exponential and $v_a(t) = e^{i\omega(t+a)}$ a time-shifted version of it.

$H[v_a](t) = e^{i\omega a} H[v](t)$ by linearity with respect to the constant $e^{i\omega a}$.

$H[v_a](t) = H[v](t+a)$ by time invariance of H.

So $H[v](t+a) = e^{i\omega a} H[v](t)$. Setting $t = 0$ and renaming we get:

$H[v](\tau) = e^{i\omega \tau} H[v](0)$

i.e. that a complex exponential $e^{i\omega \tau}$ as input will give a complex exponential of same frequency as output.

Fourier and Laplace Transforms

The eigenfunction property of exponentials is very useful for both analysis and insight into LTI systems. The Laplace transform,

$$H(s) \stackrel{\text{def}}{=} \mathcal{L}\{h(t)\} \stackrel{\text{def}}{=} \int_{-\infty}^{\infty} h(t)e^{-st}\,\mathrm{d}t$$

is exactly the way to get the eigenvalues from the impulse response. Of particular interest are pure sinusoids (i.e., exponential functions of the form $e^{j\omega t}$ where $\omega \in \mathbb{R}$ and $j \stackrel{\text{def}}{=} \sqrt{-1}$).). These are generally called complex exponentials even though the argument is purely imaginary. The Fourier transform $H(j\omega) = \mathcal{F}\{h(t)\}$ gives the eigenvalues for pure complex sinusoids. Both of $H(s)$ and $H(j\omega)$ are called the system function, system response, or transfer function.

The Laplace transform is usually used in the context of one-sided signals, i.e. signals that are zero for all values of t less than some value. Usually, this "start time" is set to zero, for convenience and without loss of generality, with the transform integral being taken from zero to infinity (the transform shown above with lower limit of integration of negative infinity is formally known as the bilateral Laplace transform).

The Fourier transform is used for analyzing systems that process signals that are infinite in extent, such as modulated sinusoids, even though it cannot be directly applied to input and output signals that are not square integrable. The Laplace transform actually works directly for these signals if they are zero before a start time, even if they are not square integrable, for stable systems. The Fourier transform is often applied to spectra of infinite signals via the Wiener–Khinchin theorem even when Fourier transforms of the signals do not exist.

Due to the convolution property of both of these transforms, the convolution that gives the output of the system can be transformed to a multiplication in the transform domain, given signals for which the transforms exist,

$$y(t) = (h * x)(t) \stackrel{\text{def}}{=} \int_{-\infty}^{\infty} h(t-\tau)x(\tau)\,\mathrm{d}\tau \stackrel{\text{def}}{=} \mathcal{L}^{-1}\{H(s)X(s)\}.$$

Not only is it often easier to do the transforms, multiplication, and inverse transform than the original convolution, but one can also gain insight into the behavior of the system from the system response. One can look at the modulus of the system function $|H(s)|$ to see whether the input $\exp(st)$ is *passed* (let through) the system or *rejected* or *attenuated* by the system (not let through).

Examples:

- A simple example of an LTI operator is the derivative.
    - $\frac{\mathrm{d}}{\mathrm{d}t}\left(c_1x_1(t)+c_2x_2(t)\right)=c_1x_1'(t)+c_2x_2'(t)$ (i.e., it is linear).
    - $\frac{\mathrm{d}}{\mathrm{d}t}x(t-\tau)=x'(t-\tau)$ (i.e., it is time invariant).

When the Laplace transform of the derivative is taken, it transforms to a simple multiplication by the Laplace variable s.

$$\mathcal{L}\left\{\frac{\mathrm{d}}{\mathrm{d}t}x(t)\right\}=sX(s)$$

That the derivative has such a simple Laplace transform partly explains the utility of the transform.

- Another simple LTI operator is an averaging operator,

$$\mathcal{A}\{x(t)\}\stackrel{\text{def}}{=}\int_{t-a}^{t+a}x(\lambda)\mathrm{d}\lambda.$$

By the linearity of integration,

$$\begin{aligned}\mathcal{A}\{c_1x_1(t)+c_2x_2(t)\}&=\int_{t-a}^{t+a}(c_1x_1(\lambda)+c_2x_2(\lambda))\mathrm{d}\lambda\\&=c_1\int_{t-a}^{t+a}x_1(\lambda)\mathrm{d}\lambda+c_2\int_{t-a}^{t-a}x_2(\lambda)\mathrm{d}\lambda\\&=c_1\mathcal{A}\{x_1(t)\}+c_2\mathcal{A}\{x_2(t)\},\end{aligned}$$

it is linear. Additionally, because

$$\begin{aligned}\mathcal{A}\{x(t-\tau)\}&=\int_{t-a}^{t+a}x(\lambda-\tau)\mathrm{d}\lambda\\&=\int_{(t-\tau)-a}^{(t-\tau)+a}x(\xi)\mathrm{d}\xi\\&=\mathcal{A}\{x\}(t-\tau),\end{aligned}$$

it is time invariant. In fact, $\mathcal{A}$ can be written as a convolution with the boxcar function $\Pi(t)$. That is,

$$\mathcal{A}\{x(t)\}=\int_{-\infty}^{\infty}\Pi\left(\frac{\lambda-t}{2a}\right)x(\lambda)\mathrm{d}\lambda,$$

where the boxcar function,

$$\Pi(t)\stackrel{\text{def}}{=}\begin{cases}1 & \text{if } |t|<\frac{1}{2},\\ 0 & \text{if } |t|>\frac{1}{2}.\end{cases}$$

Important System Properties

Some of the most important properties of a system are causality and stability. Causality is a necessity if the independent variable is time, but not all systems have time as an independent variable. For example, a system that processes still images does not need to be causal. Non-causal systems can be built and can be useful in many circumstances. Even non-real systems can be built and are very useful in many contexts.

Causality

A system is causal if the output depends only on present and past, but not future inputs. A necessary and sufficient condition for causality is:

$$h(t) = 0 \quad \forall t < 0,$$

where $h(t)$ is the impulse response. It is not possible in general to determine causality from the Laplace transform, because the inverse transform is not unique. When a region of convergence is specified, then causality can be determined.

Stability

A system is bounded-input, bounded-output stable (BIBO stable) if, for every bounded input, the output is finite. Mathematically, if every input satisfying:

$$\| x(t) \|_{\infty} < \infty$$

leads to an output satisfying:

$$\| y(t) \|_{\infty} < \infty$$

(that is, a finite maximum absolute value of $x(t)$ implies a finite maximum absolute value of $y(t)$), then the system is stable. A necessary and sufficient condition is that $h(t)$, the impulse response, is in L^1 (has a finite L^1 norm):

$$\| h(t) \|_1 = \int_{-\infty}^{\infty} |h(t)| \mathrm{d}t < \infty.$$

In the frequency domain, the region of convergence must contain the imaginary axis $s = j\omega.$.

As an example, the ideal low-pass filter with impulse response equal to a sinc function is not BIBO stable, because the sinc function does not have a finite L^1 norm. Thus, for some bounded input, the output of the ideal low-pass filter is unbounded. In particular, if the input is zero for $t < 0$ and equal to a sinusoid at the cut-off frequency for $t > 0$, then the output will be unbounded for all times other than the zero crossings.

Discrete-time Systems

Almost everything in continuous-time systems has a counterpart in discrete-time systems.

Discrete-time Systems from Continuous-time Systems

In many contexts, a discrete time (DT) system is really part of a larger continuous time (CT) system. For example, a digital recording system takes an analog sound, digitizes it, possibly processes the digital signals, and plays back an analog sound for people to listen to.

Formally, the DT signals studied are almost always uniformly sampled versions of CT signals. If $x(t)$ is a CT signal, then an analog to digital converter will transform it to the DT signal:

$$x[n] \stackrel{\text{def}}{=} x(nT) \qquad \forall n \in \mathbb{Z},$$

where T is the sampling period. It is very important to limit the range of frequencies in the input signal for faithful representation in the DT signal, since then the sampling theorem guarantees that no information about the CT signal is lost. A DT signal can only contain a frequency range of $1/(2T)$; other frequencies are aliased to the same range.

Impulse Response and Convolution

Let $\{x[m-k];\ m\}$ represent the sequence $\{x[m-k];\ \text{for all integer values of } m\}$.

And let the shorter notation $\{x\}$ represent $\{x[m];\ m\}$.

A discrete system transforms an input sequence, $\{x\}$ into an output sequence, $\{y\}$. In general, every element of the output can depend on every element of the input. Representing the transformation operator by O, we can write:

$$y[n] \stackrel{\text{def}}{=} O_n\{x\}.$$

Note that unless the transform itself changes with n, the output sequence is just constant, and the system is uninteresting. (Thus the subscript, n.) In a typical system, y[n] depends most heavily on the elements of x whose indices are near n.

For the special case of the Kronecker delta function, $x[m] = \delta[m]$, the output sequence is the impulse response:

$$h[n] \stackrel{\text{def}}{=} O_n\{\delta[m];\ m\}.$$

For a linear system, O must satisfy:

$$O_n\left\{\sum_{k=-\infty}^{\infty} c_k \cdot x_k[m];\ m\right\} = \sum_{k=-\infty}^{\infty} c_k \cdot O_n\{x_k\}.$$

And the time-invariance requirement is:

$$\begin{aligned} O_n\{x[m-k];\ m\} &= y[n-k] \\ &\stackrel{\text{def}}{=} O_{n-k}\{x\}. \end{aligned}$$

In such a system, the impulse response, $\{h\}$ characterizes the system completely. i.e., for any input sequence, the output sequence can be calculated in terms of the input and the impulse response. To see how that is done, consider the identity:

$$x[m] \equiv \sum_{k=-\infty}^{\infty} x[k]\cdot\delta[m-k],$$

which expresses $\{x\}$ in terms of a sum of weighted delta functions.

Therefore:

$$y[n] = O_n\{x\} = O_n\left\{\sum_{k=-\infty}^{\infty} x[k]\cdot\delta[m-k];\, m\right\}$$

$$= \sum_{k=-\infty}^{\infty} x[k]\cdot O_n\{\delta[m-k];\, m\},$$

where we have invoked eq. $O_n\left\{\sum_{k=-\infty}^{\infty} c_k\cdot x_k[m];\, m\right\} = \sum_{k=-\infty}^{\infty} c_k\cdot O_n\{x_k\}$ for the case $c_k = x[k]$ and $x_k[m] = \delta[m-k]$.

And because of eq. $O_n\{x[m-k];\, m\} = y[n-k] \stackrel{\text{def}}{=} O_{n-k}\{x\}$ we may write:

$$O_n\{\delta[m-k];\, m\} = O_{n-k}\{\delta[m];\, m\}$$

$$\stackrel{\text{def}}{=} h[n-k].$$

Therefore:

$$y[n] = \sum_{k=-\infty}^{\infty} x[k]\cdot h[n-k]$$

$$= \sum_{k=-\infty}^{\infty} x[n-k]\cdot h[k], \text{ (commutativity)}$$

which is the familiar discrete convolution formula. The operator O_n can therefore be interpreted as proportional to a weighted average of the function x[k]. The weighting function is h[-k], simply shifted by amount n. As n changes, the weighting function emphasizes different parts of the input function. Equivalently, the system's response to an impulse at n=0 is a “time” reversed copy of the unshifted weighting function. When h[k] is zero for all negative k, the system is said to be causal.

Exponentials as Eigenfunctions

An eigenfunction is a function for which the output of the operator is the same function, scaled by some constant. In symbols,

$$\mathcal{H}f = \lambda f,$$

where f is the eigenfunction and λ is the eigenvalue, a constant.

The exponential functions $z^n = e^{sTn}$, where $n \in \mathbb{Z}$, are eigenfunctions of a linear, time-invariant operator. $T \in \mathbb{R}$ is the sampling interval, and $z = e^{sT}$, $z, s \in \mathbb{C}$. A simple proof illustrates this concept.

Suppose the input is $h[n]$. The output of the system with impulse response $h[n]$ is then,

$$\sum_{m=-\infty}^{\infty} h[n-m]z^m$$

which is equivalent to the following by the commutative property of convolution,

$$\sum_{m=-\infty}^{\infty} h[m]z^{(n-m)} = z^n \sum_{m=-\infty}^{\infty} h[m]z^{-m} = z^n H(z)$$

where,

$$H(z) \stackrel{\text{def}}{=} \sum_{m=-\infty}^{\infty} h[m]z^{-m}$$

is dependent only on the parameter z.

So z^n is an eigenfunction of an LTI system because the system response is the same as the input times the constant .

Z and Discrete-time Fourier Transforms

The eigenfunction property of exponentials is very useful for both analysis and insight into LTI systems. The Z transfor,

$$H(z) = \mathcal{Z}\{h[n]\} = \sum_{n=-\infty}^{\infty} h[n]z^{-n}$$

is exactly the way to get the eigenvalues from the impulse response. Of particular interest are pure sinusoids, i.e. exponentials of the form $e^{j\omega n}$, where $\omega \in \mathbb{R}$. These can also be written as z^n with $z = e^{j\omega}$. These are generally called complex exponentials even though the argument is purely imaginary. The discrete-time Fourier transform (DTFT) $H(e^{j\omega}) = \mathcal{F}\{h[n]\}$ gives the eigenvalues of pure sinusoids. Both of $H(z)$ and $H(e^{j\omega})$ are called the system function, system response, or transfer function'.

The Z transform is usually used in the context of one-sided signals, i.e. signals that are zero for all values of t less than some value. Usually, this "start time" is set to zero, for convenience and without loss of generality. The Fourier transform is used for analyzing signals that are infinite in extent.

Due to the convolution property of both of these transforms, the convolution that gives the output of the system can be transformed to a multiplication in the transform domain. That is,

$$[n] = (h * x)[n] = \sum_{m=-\infty}^{\infty} h[n-m]x[m] = \mathcal{Z}^{-1}\{H(z)X(z)\}.$$

Just as with the Laplace transform transfer function in continuous-time system analysis, the Z transform makes it easier to analyze systems and gain insight into their behavior. One can look at the modulus of the system function $|H(z)|$ to see whether the input z^n is *passed* (let through) by the system, or *rejected* or *attenuated* by the system (not let through).

Examples:

- A simple example of an LTI operator is the delay operator $D\{x[n]\} \overset{\text{def}}{=} x[n-1]$.
    - $D\left(c_1 \cdot x_1[n] + c_2 \cdot x_2[n]\right) = c_1 \cdot x_1[n-1] + c_2 \cdot x_2[n-1] = c_1 \cdot Dx_1[n] + c_2 \cdot Dx_2[n]$ (i.e., it is linear).
    - $D\{x[n-m]\} = x[n-m-1] = x[(n-1)-m] = D\{x\}[n-m]$ (i.e., it is time invariant).

The Z transform of the delay operator is a simple multiplication by z−1. That is,

$$\mathcal{Z}\{Dx[n]\} = z^{-1}X(z).$$

- Another simple LTI operator is the averaging operator,

$$\mathcal{A}\{x[n]\} \overset{\text{def}}{=} \sum_{k=n-a}^{n+a} x[k].$$

Because of the linearity of sums,

$$\begin{aligned}\mathcal{A}\{c_1x_1[n] + c_2x_2[n]\} &= \sum_{k=n-a}^{n+a} \left(c_1x_1[k] + c_2x_2[k]\right) \\ &= c_1 \sum_{k=n-a}^{n+a} x_1[k] + c_2 \sum_{k=n-a}^{n+a} x_2[k] \\ &= c_1\mathcal{A}\{x_1[n]\} + c_2\mathcal{A}\{x_2[n]\},\end{aligned}$$

and so it is linear. Because,

$$\begin{aligned}\mathcal{A}\{x[n-m]\} &= \sum_{k=n-a}^{n+a} x[k-m] \\ &= \sum_{k'=(n-m)-a}^{(n-m)+a} x[k'] \\ &= \mathcal{A}\{x\}[n-m],\end{aligned}$$

it is also time invariant.

Important System Properties

The input-output characteristics of discrete-time LTI system are completely described by its impulse response $h[n]$. Some of the most important properties of a system are causality and stability. Unlike CT systems, non-causal DT systems can be realized. It is trivial to make an acausal FIR system causal by adding delays. It is even possible to make acausal IIR systems. Non-stable systems can be built and can be useful in many circumstances. Even non-real systems can be built and are very useful in many contexts.

Causality

A discrete-time LTI system is causal if the current value of the output depends on only the current value and past values of the input., A necessary and sufficient condition for causality is,

$$h[n]=0\ \forall n<0,$$

where $h[n]$ is the impulse response. It is not possible in general to determine causality from the Z transform, because the inverse transform is not unique. When a region of convergence is specified, then causality can be determined.

Stability

A system is bounded input, bounded output stable (BIBO stable) if, for every bounded input, the output is finite. Mathematically, if

$$\|x[n]\|_{\infty}<\infty$$

implies that

$$\|y[n]\|_{\infty}<\infty$$

(that is, if bounded input implies bounded output, in the sense that the maximum absolute values of $x[n]$ and $y[n]$ are finite), then the system is stable. A necessary and sufficient condition is that $h[n]$, the impulse response, satisfies

$$\|h[n]\|_1 \stackrel{\text{def}}{=} \sum_{n=-\infty}^{\infty}|h[n]|<\infty.$$

In the frequency domain, the region of convergence must contain the unit circle (i.e., the locus satisfying $|z|=1$ for complex z).

Causal System

In control theory, a causal system (also known as a physical or nonanticipative system) is a system where the output depends on past and current inputs but not future inputs—i.e., the output $y(t_0)$ depends on only the input $x(t)$ for values of $t<t_0$.

The idea that the output of a function at any time depends only on past and present values of input is defined by the property commonly referred to as causality. A system that has *some* dependence on input values from the future (in addition to possible dependence on past or current input values) is termed a non-causal or acausal system, and a system that depends *solely* on future input values is an anticausal system.

Classically, nature or physical reality has been considered to be a causal system. Physics involving special relativity or general relativity require more careful definitions of causality, as described elaborately in Causality (physics).

The causality of systems also plays an important role in digital signal processing, where filters are constructed so that they are causal, sometimes by altering a non-causal formulation to remove the lack of causality so that it is realizable.

For a causal system, the impulse response of the system must use only the present and past values of the input to determine the output. This requirement is a necessary and sufficient condition for a system to be causal, regardless of linearity. Note that similar rules apply to either discrete or continuous cases. By this definition of requiring no future input values, systems must be causal to process signals in real time.

Mathematical Definitions

Definition 1: A system mapping x to y is causal if and only if, for any pair of input signals $x_2(t)$, $x_2(t)$ and any choice of t_0, such that,

$$x_1(t) = x_2(t), \quad \forall\, t < t_0,$$

the corresponding outputs satisfy,

$$y_1(t) = y_2(t), \quad \forall\, t < t_0$$

Definition 2: Suppose $h(t)$ is the impulse response of any system H described by a linear constant coefficient differential equation. The system H is causal if and only if,

$$h(t) = 0, \quad \forall\, t < 0$$

otherwise it is non-causal.

Examples:

The following examples are for systems with an input x and output y.

Examples of causal systems:

- Memoryless system,

  $$y(t) = 1 - x(t)\cos(\omega t)$$

- Autoregressive filter,

  $$y(t) = \int_0^{\infty} x(t-\tau)e^{-\beta\tau}\, d\tau$$

Examples of non-causal (acausal) systems:

$$y(t) = \int_{-\infty}^{\infty} \sin(t+\tau)x(\tau)d\tau$$

- Central moving average,

$$y_n = \frac{1}{2}x_{n-1} + \frac{1}{2}x_{n+1}$$

Examples of anti-causal systems:

$$y(t) = \int_0^{\infty} \sin(t+\tau)d\tau$$

- Look-ahead.

$$y_n = x_{n+1}$$

Difference Equations

One of the most important concepts of DSP is to be able to properly represent the input/output relationship to a given LTI system. A linear constant-coefficient difference equation (LCCDE) serves as a way to express just this relationship in a discrete-time system. Writing the sequence of inputs and outputs, which represent the characteristics of the LTI system, as a difference equation help in understanding and manipulating a system.

An equation that shows the relationship between consecutive values of a sequence and the differences among them. They are often rearranged as a recursive formula so that a systems output can be computed from the input signal and past outputs.

Example:

$$y[n] + 7y[n-1] + 2y[n-2] = x[n] - 4x[n-1]$$

General Formulas for the Difference Equation

As stated briefly in the definition above, a difference equation is a very useful tool in describing and calculating the output of the system described by the formula for a given sample n. The key property of the difference equation is its ability to help easily find the transform, $H(z)$, of a system. In the following two subsections, we will look at the general form of the difference equation and the general conversion to a z-transform directly from the difference equation.

The general form of a linear, constant-coefficient difference equation (LCCDE), is shown below:

$$\sum_{k=0}^{N} a_k y[n-k] = \sum_{k=0}^{N} b_k x[n-k]$$

We can also write the general form to easily express a recursive output, which looks like this:

$$y[n] = -\sum_{k=1}^{N} a_k y[n-k] + \sum_{k=0}^{N} b_k x[n-k]$$

From this equation, note that $y[n-k]$ represents the outputs and $x[n-k]$ represents the inputs. The value of N represents the order of the difference equation and corresponds to the memory of the system being represented. Because this equation relies on past values of the output, in order to compute a numerical solution, certain past outputs, referred to as the initial conditions, must be known.

Conversion to Z-transform

Using the above formula, eq. $\sum_{k=0}^{N} a_k y[n-k] = \sum_{k=0}^{N} b_k x[n-k]$, we can easily generalize the transfer function, $H(z)$, for any difference equation. Below are the steps taken to convert any difference equation into its transfer function, i.e. z-transform. The first step involves taking the Fourier Transform of all the terms in Equation. Then we use the linearity property to pull the transform inside the summation and the time-shifting property of the z-transform to change the time-shifting terms to exponentials. Once this is done, we arrive at the following equation: $a_0 = 1$.

$$Y[z] = -\sum_{k=1}^{N} a_k Y(z) z^{-k} + \sum_{k=0}^{N} b_k X(z) z^{-k}$$

$$H(z) = \frac{Y(z)}{X(z)}$$

$$= \frac{\sum_{k=0}^{M} b_k z^{-k}}{1 + \sum_{k=1}^{N} a_k z^{-k}}$$

Conversion to Frequency Response

Once the z-transform has been calculated from the difference equation, we can go one step further to define the frequency response of the system, or filter, that is being represented by the difference equation.

Below is the general formula for the frequency response of a z-transform. The conversion is simple a matter of taking the z-transform formula, $H(z)$, and replacing every instance of z with e^{iw}.

$$H(w) = H(z)|_{z,z=eiw}$$

$$= \frac{\sum_{k=0}^{M} b_k e^{-(iwk)}}{\sum_{k=0}^{N} a_k e^{-(iwk)}}$$

Once you understand the derivation of this formula, look at the module concerning Filter Design from the Z-Transform for a look into how all of these ideas of the Z-transform, Difference Equation, and Pole/Zero Plots play a role in filter design.

Example:

Finding Difference Equation

Below is a basic example showing the opposite of the steps above: given a transfer function one can easily calculate the systems difference equation.

$$H(z)=\frac{(z+1)^2}{\left(z-\frac{1}{2}\right)\left(z+\frac{3}{4}\right)}$$

Given this transfer function of a time-domain filter, we want to find the difference eq. $\sum_{k=0}^{N} a_k y[n-k]=\sum_{k=0}^{N} b_k x[n-k]$. To begin with, expand both polynomials and divide them by the highest order z.

$$H(z)=\frac{(z+1)(z+1)}{\left(z-\frac{1}{2}\right)\left(z+\frac{3}{4}\right)}$$
$$=\frac{z^2+2z+1}{z^2+2z+1-\frac{3}{8}}$$
$$=\frac{1+2z^{-1}+z^{-2}}{1+\frac{1}{4}z^{-1}-\frac{3}{8}z^{-2}}$$

From this transfer function, the coefficients of the two polynomials will be our a_k and b_k values found in the general difference equation formula, Equation. Using these coefficients and the above form of the transfer function, we can easily write the difference equation:

$$x[n]+2x[n-1]+x[n-2]=y[n]+\frac{1}{4}y[n-1]-\frac{3}{8}y[n-2]$$

$$y[n]+x[n]+2x[n-1]+x[n-2]+\frac{-1}{4}y[n-1]+\frac{3}{8}y[n-2]$$

Solving a LCCDE

In order for a linear constant-coefficient difference equation to be useful in analyzing a LTI system, we must be able to find the systems output based upon a known input, $x(n)$, and a set of initial conditions. Two common methods exist for solving a LCCDE: the direct method and the indirect method, the later being based on the z-transform.

Direct Method

The final solution to the output based on the direct method is the sum of two parts, expressed in the following equation:

$$y(n)=y_h(n)+y_p(n)$$

The first part, $y_h(n)$, is referred to as the homogeneous solution and the second part, $y_h(n)$, is referred to as particular solution. The following method is very similar to that used to solve many differential equations, so if you have taken a differential calculus course or used differential equations before then this should seem very familiar.

Homogeneous Solution

We begin by assuming that the input is zero, $x(n)=0$. Now we simply need to solve the homogeneous difference equation:

$$\sum_{k=0}^{N} a_k y[n-k]=0$$

In order to solve this, we will make the assumption that the solution is in the form of an exponential. We will use lambda, λ, to represent our exponential terms. We now have to solve the following equation:

$$\sum_{k=0}^{N} a_k \lambda^{n-k}=0$$

We can expand this equation out and factor out all of the lambda terms. This will give us a large polynomial in parenthesis, which is referred to as the characteristic polynomial. The roots of this polynomial will be the key to solving the homogeneous equation. If there are all distinct roots, then the general solution to the equation will be as follows:

$$y_h(n)=C_1(\lambda_1)^n+C_2(\lambda_2)^n+\ldots+C_N(\lambda_N)^n$$

However, if the characteristic equation contains multiple roots then the above general solution will be slightly different. Below we have the modified version for an equation where λ_1 has K multiple roots:

$$y_h(n)=C_1(\lambda_1)^n+C_1n(\lambda_1)^n+C_1n^2(\lambda_1)^n+\ldots+C_1n^{K-1}(\lambda_1)^n+C_2(\lambda_2)^n+\ldots+C_N(\lambda_N)^n$$

Particular Solution

The particular solution, $y_p(n)$, will be any solution that will solve the general difference equation:

$$\sum_{k=0}^{N} a_k y_p(n-k)=\sum_{k=0}^{M} b_k x(n-k)$$

In order to solve, our guess for the solution to $y_p(n)$ will take on the form of the input, $x(n)$. After guessing at a solution to the above equation involving the particular solution, one only needs to plug the solution into the difference equation and solve it out.

Indirect Method

The indirect method utilizes the relationship between the difference equation and z-transform, to find a solution. The basic idea is to convert the difference equation into a z-transform, as described above, to get the resulting output, $Y(z)$. Then by inverse transforming this and using partial-fraction expansion, we can arrive at the solution.

$$Z\{y(n+1)-y(n)\}=zY(z)-y(0)$$

This can be interatively extended to an arbitrary order derivative as in Equation,

$$Z\left\{\sum_{m=0}^{N-1} y(n-m)\right\} = z^n Y(z) - \sum_{m=0}^{N-1} z^{n-m-1} \mathrm{y}^{(m)}(0)$$

Now, the Laplace transform of each side of the differential equation can be taken,

$$Z\left\{\sum_{k=0}^{N} a_k \left[y(n-m+1) - \sum_{k=0}^{N} y(n-m)y(n) \right] = Z\{x(n)\}\right\}$$

which by linearity results in,

$$\sum_{k=0}^{N} a_k Z\left\{ y(n-m+1) - \sum_{m=0}^{N-1} y(n-m)y(n) \right\} = Z\{x(n)\}$$

and by differentiation properties in,

$$\sum_{k=0}^{N} a_k \left\{ z^k Z\{y(n)\} - \sum_{m=0}^{N-1} z^{k-m-1} y^{(m)}(0) \right\} = Z\{x(n)\}.$$

Rearranging terms to isolate the Laplace transform of the output,

$$Z\{y(n)\} = \frac{Z\{x(n)\} + \sum_{k=0}^{N} \sum_{m=0}^{k-1} a_k z^{k-m-1} y^{(m)}(0)}{\sum_{k=0}^{N} a_k z^k}.$$

Thus, it is found that,

$$Y(\mathrm{z}) = \frac{X(z)\} + \sum_{k=0}^{N} \sum_{m=0}^{k-1} a_k z^{k-m-1} y^{(m)}(0)}{\sum_{k=0}^{N} a_k z^k}.$$

In order to find the output, it only remains to find the Laplace transform $X(z)$ of the input, substitute the initial conditions, and compute the inverse Z-transform of the result. Partial fraction expansions are often required for this last step. This may sound daunting while looking at equation above, but it is often easy in practice, especially for low order difference equations above. Equation can also be used to determine the transfer function and frequency response.

As an example, consider the difference equation,

$$y[n-2] + 4y[n-1] + 3y[n] = cos(n)$$

with the initial conditions $y'(0) = 1$ and $y(0) = 0$ Using the method described above, the Z transform of the solution $y[n]$ is given by,

$$Y[z] = \frac{z}{[z^2+1][z+1][z+3]} + \frac{1}{[z+1][z+3]}.$$

Performing a partial fraction decomposition, this also equals

$$Y[z] = .25\frac{1}{z+1} - .35\frac{1}{z+3} + .1\frac{z}{z^2+1} + .2\frac{1}{z^2+1}.$$

Computing the inverse Laplace transform,

$$y(n) = (.25z^{-n} - .35z^{-3n} + .1\cos(n) + .2\sin(n))u(n)$$

One can check that this satisfies that this satisfies both the differential equation and the initial conditions.

Impulse Response

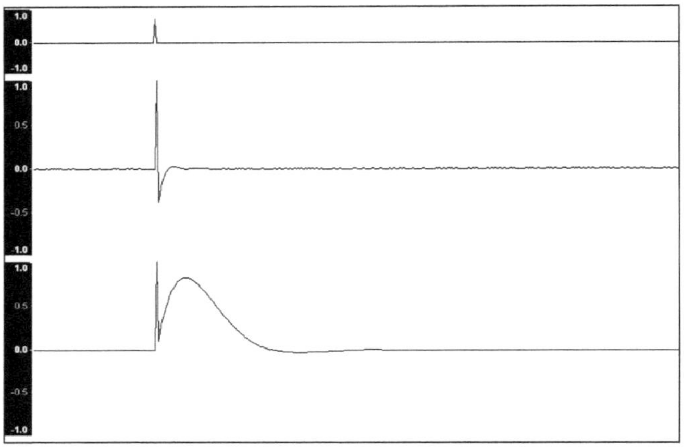

The Impulse response from a simple audio system. Showing, from top to bottom, the original impulse, the response after high frequency boosting, and the response after low frequency boosting.

In signal processing, the impulse response, or impulse response function (IRF), of a dynamic system is its output when presented with a brief input signal, called an impulse. More generally, an impulse response is the reaction of any dynamic system in response to some external change. In both cases, the impulse response describes the reaction of the system as a function of time (or possibly as a function of some other independent variable that parameterizes the dynamic behavior of the system).

In all these cases, the dynamic system and its impulse response may be actual physical objects, or may be mathematical systems of equations describing such objects. Since the impulse function contains all frequencies, the impulse response defines the response of a linear time-invariant system for all frequencies.

Mathematical Considerations

Mathematically, how the impulse is described depends on whether the system is modeled in discrete or continuous time. The impulse can be modeled as a Dirac delta function for continuous-time systems, or as the Kronecker delta for discrete-time systems. The Dirac delta represents the limiting case of a pulse made very short in time while maintaining its area or integral (thus giving an infinitely high peak). While this is impossible in any real system, it is a useful idealisation.

In Fourier analysis theory, such an impulse comprises equal portions of all possible excitation frequencies, which makes it a convenient test probe.

Any system in a large class known as *linear, time-invariant* (LTI) is completely characterized by its impulse response. That is, for any input, the output can be calculated in terms of the input and the impulse response. The impulse response of a linear transformation is the image of Dirac's delta function under the transformation, analogous to the fundamental solution of a partial differential operator.

It is usually easier to analyze systems using transfer functions as opposed to impulse responses. The transfer function is the Laplace transform of the impulse response. The Laplace transform of a system's output may be determined by the multiplication of the transfer function with the input's Laplace transform in the complex plane, also known as the frequency domain. An inverse Laplace transform of this result will yield the output in the time domain.

To determine an output directly in the time domain requires the convolution of the input with the impulse response. When the transfer function and the Laplace transform of the input are known, this convolution may be more complicated than the alternative of multiplying two functions in the frequency domain. The impulse response, considered as a Green's function, can be thought of as an "influence function": how a point of input influences output.

Practical Applications

In practical systems, it is not possible to produce a perfect impulse to serve as input for testing; therefore, a brief pulse is sometimes used as an approximation of an impulse. Provided that the pulse is short enough compared to the impulse response, the result will be close to the true, theoretical, impulse response. In many systems, however, driving with a very short strong pulse may drive the system into a nonlinear regime, so instead the system is driven with a pseudo-random sequence, and the impulse response is computed from the input and output signals.

Loudspeakers

An application that demonstrates this idea was the development of impulse response loudspeaker testing in the 1970s. Loudspeakers suffer from phase inaccuracy, a defect unlike other measured properties such as frequency response. Phase inaccuracy is caused by (slightly) delayed frequencies/octaves that are mainly the result of passive cross overs (especially higher order filters) but are also caused by resonance, energy storage in the cone, the internal volume, or the enclosure panels vibrating. Measuring the impulse response, which is a direct plot of this "time-smearing," provided a tool for use in reducing resonances by the use of improved materials for cones and enclosures, as well as changes to the speaker crossover. The need to limit input amplitude to maintain the linearity of the system led to the use of inputs such as pseudo-random maximum length sequences, and to the use of computer processing to derive the impulse response.

Electronic Processing

Impulse response analysis is a major facet of radar, ultrasound imaging, and many areas of digital signal processing. An interesting example would be broadband internet connections.

DSL/Broadband services use adaptive equalisation techniques to help compensate for signal distortion and interference introduced by the copper phone lines used to deliver the service.

Control Systems

In control theory the impulse response is the response of a system to a Dirac delta input. This proves useful in the analysis of dynamic systems; the Laplace transform of the delta function is 1, so the impulse response is equivalent to the inverse Laplace transform of the system's transfer function.

Acoustic and Audio Applications

In acoustic and audio applications, impulse responses enable the acoustic characteristics of a location, such as a concert hall, to be captured. Various packages are available containing impulse responses from specific locations, ranging from small rooms to large concert halls. These impulse responses can then be utilized in convolution reverb applications to enable the acoustic characteristics of a particular location to be applied to target audio.

Economics

In economics, and especially in contemporary macroeconomic modeling, impulse response functions are used to describe how the economy reacts over time to exogenous impulses, which economists usually call shocks, and are often modeled in the context of a vector autoregression. Impulses that are often treated as exogenous from a macroeconomic point of view include changes in government spending, tax rates, and other fiscal policy parameters; changes in the monetary base or other monetary policy parameters; changes in productivity or other technological parameters; and changes in preferences, such as the degree of impatience. Impulse response functions describe the reaction of endogenous macroeconomic variables such as output, consumption, investment, and employment at the time of the shock and over subsequent points in time. Recently, asymmetric impulse response functions have been suggested in the literature that separate the impact of a positive shock from a negative one.

Step Response

The step response of a system in a given initial state consists of the time evolution of its outputs when its control inputs are Heaviside step functions. In electronic engineering and control theory, step response is the time behaviour of the outputs of a general system when its inputs change from zero to one in a very short time. The concept can be extended to the abstract mathematical notion of a dynamical system using an evolution parameter.

From a practical standpoint, knowing how the system responds to a sudden input is important because large and possibly fast deviations from the long term steady state may have extreme effects on the component itself and on other portions of the overall system dependent on this component. In addition, the overall system cannot act until the component's output settles down to some vicinity of its final state, delaying the overall system response. Formally, knowing the step response of a dynamical system gives information on the stability of such a system, and on its ability to reach one stationary state when starting from another.

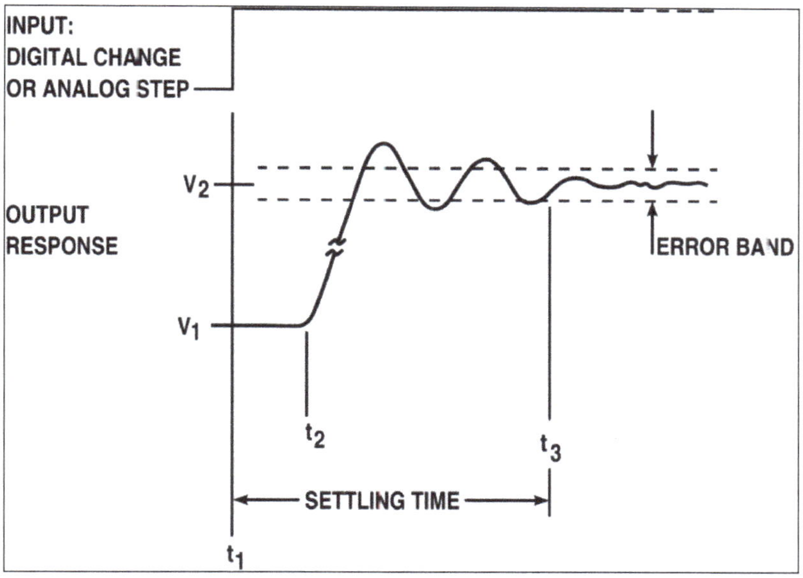

A typical step response for a second order system, illustrating overshoot, followed by ringing, all subsiding within a settling time.

Time Domain versus Frequency Domain

Instead of frequency response, system performance may be specified in terms of parameters describing time-dependence of response. The step response can be described by the following quantities related to its time behavior,

- Overshoot,
- Rise time,
- Settling time,
- Ringing.

In the case of linear dynamic systems, much can be inferred about the system from these characteristics. Below the step response of a simple two-pole amplifier is presented, and some of these terms are illustrated.

Step Response of Feedback Amplifiers

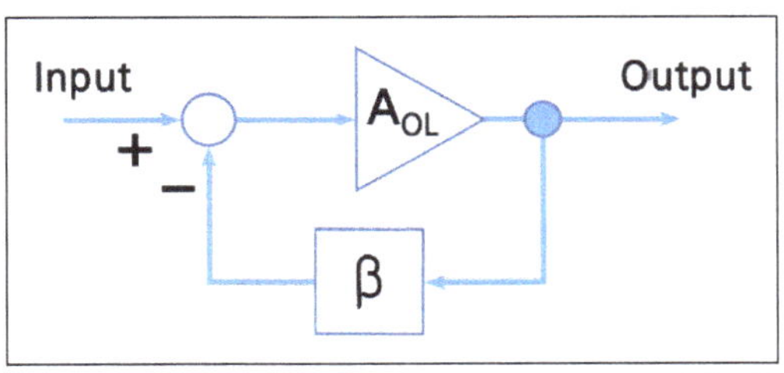

Ideal negative feedback model; open loop gain is A_{OL} and feedback factor is β.

This section describes the step response of a simple negative feedback amplifier shown in figure. The feedback amplifier consists of a main open-loop amplifier of gain A_{OL} and a feedback loop governed by a feedback factor β. This feedback amplifier is analyzed to determine how its step response depends upon the time constants governing the response of the main amplifier, and upon the amount of feedback used.

A negative-feedback amplifier has gain given by:

$${}_{FB} = \frac{A_{OL}}{1+\beta A_{OL}},$$

where A_{OL} = open-loop gain, A_{FB} = closed-loop gain (the gain with negative feedback present) and β = feedback factor.

With One Dominant Pole

In many cases, the forward amplifier can be sufficiently well modeled in terms of a single dominant pole of time constant τ, that it, as an open-loop gain given by:

$$A_{OL} = \frac{A_0}{1+j\omega\tau},$$

with zero-frequency gain A_0 and angular frequency $\omega = 2\pi f$. This forward amplifier has unit step response,

$$S_{OL}(t) = A_0(1-e^{-t/\tau}),$$

an exponential approach from 0 toward the new equilibrium value of A_0.

The one-pole amplifier's transfer function leads to the closed-loop gain:

$$A_{FB} = \frac{A_0}{1+\beta A_0} \bullet \frac{1}{1+j\omega\frac{\tau}{1+\beta A_0}}.$$

This closed-loop gain is of the same form as the open-loop gain: a one-pole filter. Its step response is of the same form: an exponential decay toward the new equilibrium value. But the time constant of the closed-loop step function is $\tau / (1 + \beta A_0)$, so it is faster than the forward amplifier's response by a factor of $1 + \beta A_0$:

$$S_{FB}(t) = \frac{A_0}{1+\beta A_0}(1-e^{-t(1+\beta A_0)/\tau})$$

As the feedback factor β is increased, the step response will get faster, until the original assumption of one dominant pole is no longer accurate. If there is a second pole, then as the closed-loop time constant approaches the time constant of the second pole, a two-pole analysis is needed.

Two-pole Amplifiers

In the case that the open-loop gain has two poles (two time constants, τ_1, τ_2), the step response is a bit more complicated. The open-loop gain is given by:

$$A_{OL} = \frac{A_0}{(1 + j\omega\tau_1)(1 + j\omega\tau_2)},$$

with zero-frequency gain A_0 and angular frequency $\omega = 2\pi f$.

Analysis

The two-pole amplifier's transfer function leads to the closed-loop gain:

$$A_{FB} = \frac{A_0}{1 + \beta A_0} \bullet \frac{1}{1 - j\omega\frac{\tau_1 + \tau_2}{1 + \beta A_0} + (j\omega)^2 \frac{\tau_1\tau_2}{1 + \beta A_0}}.$$

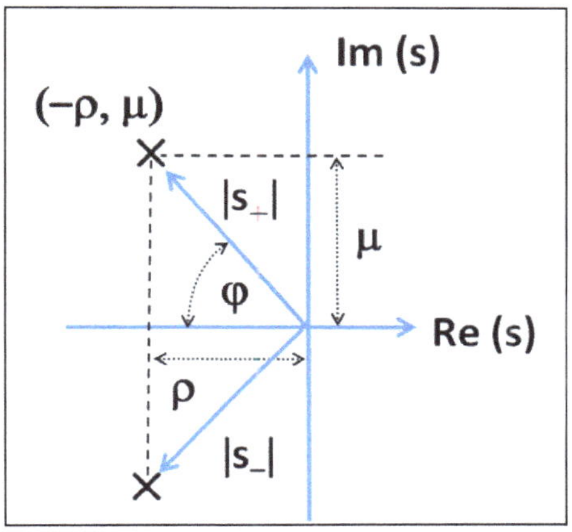

Conjugate pole locations for a two-pole feedback amplifier; *Re*(s) = real axis and *Im*(s) = imaginary axis.

The time dependence of the amplifier is easy to discover by switching variables to $s = j\omega$, whereupon the gain becomes:

$$A_{FB} = \frac{A_0}{\tau_1\tau_2} \bullet \frac{1}{s^2 + s\left(\frac{1}{\tau_1} + \frac{1}{\tau_2}\right) + \frac{1 + \beta A_0}{\tau_1\tau_2}}$$

The poles of this expression (that is, the zeros of the denominator) occur at:

$$2s = -\left(\frac{1}{\tau_1} + \frac{1}{\tau_2}\right)$$

$$\pm\sqrt{\left(\frac{1}{\tau_1} - \frac{1}{\tau_2}\right)^2 - \frac{4\beta A_0}{\tau_1\tau_2}},$$

which shows for large enough values of βA_0 the square root becomes the square root of a negative number, that is the square root becomes imaginary, and the pole positions are complex conjugate numbers, either s_+ or s_-;

$$s_{\pm} = -\rho \pm j\mu,$$

with

$$\rho = \frac{1}{2}\left(\frac{1}{\tau_1} + \frac{1}{\tau_2}\right),$$

and

$$= \frac{1}{2}\sqrt{\frac{4\beta A_0}{\tau_1\tau_2} - \left(\frac{1}{\tau_1} - \frac{1}{\tau_2}\right)^2}.$$

Using polar coordinates with the magnitude of the radius to the roots given by $|s|$:

$$|s| = |s_{\pm}| = \sqrt{\rho^2 + \mu^2},$$

and the angular coordinate φ is given by:

$$\cos\phi = \frac{\rho}{|s|}, \sin\phi = \frac{\mu}{|s|}.$$

Tables of Laplace transforms show that the time response of such a system is composed of combinations of the two functions:

$$e^{-\rho t}\sin(\mu t) \text{ and } e^{-\rho t}\cos(\mu t)$$

which is to say, the solutions are damped oscillations in time. In particular, the unit step response of the system is,

$$S(t) = \left(\frac{A_0}{1+\beta A_0}\right)\left(1 - e^{-\rho t}\frac{\sin(\mu t + \phi)}{\sin(\phi)}\right),$$

which simplifies to,

$$S(t) = 1 - e^{-\rho t}\frac{\sin(\mu t + \phi)}{\sin(\phi)}$$

when A_0 tends to infinity and the feedback factor β is one.

Notice that the damping of the response is set by ρ, that is, by the time constants of the open-loop amplifier. In contrast, the frequency of oscillation is set by μ, that is, by the feedback parameter through βA_0. Because ρ is a sum of reciprocals of time constants, it is interesting to notice that ρ is dominated by the *shorter* of the two.

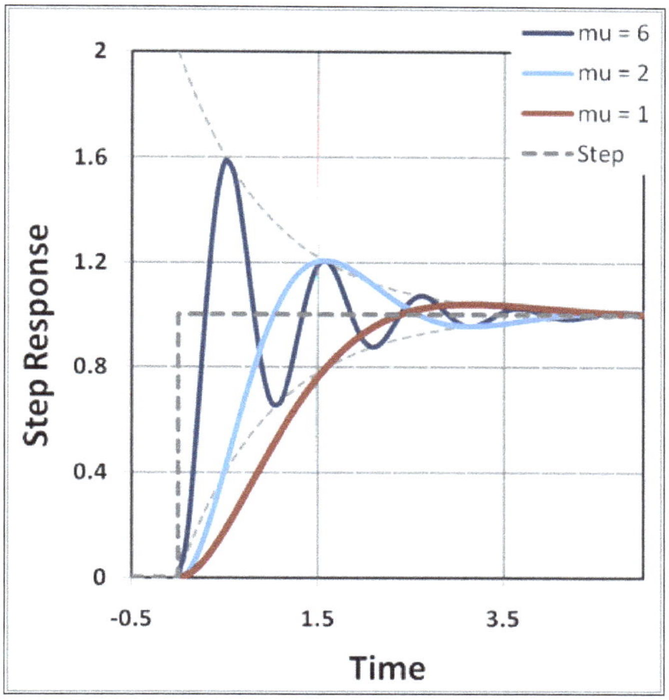

Step-response of a linear two-pole feedback amplifier; time is in units of $1/\rho$, that is, in terms of the time constants of A_{OL}; curves are plotted for three values of *mu* = μ, which is controlled by β.

The time response to a unit step input for three values of the parameter μ. It can be seen that the frequency of oscillation increases with μ, but the oscillations are contained between the two asymptotes set by the exponentials [1 – exp (−ρt)] and [1 + exp(−ρt)]. These asymptotes are determined by ρ and therefore by the time constants of the open-loop amplifier, independent of feedback.

The phenomenon of oscillation about the final value is called ringing. The overshoot is the maximum swing above final value, and clearly increases with μ. Likewise, the undershoot is the minimum swing below final value, again increasing with μ. The settling time is the time for departures from final value to sink below some specified level, say 10% of final value.

The dependence of settling time upon μ is not obvious, and the approximation of a two-pole system probably is not accurate enough to make any real-world conclusions about feedback dependence of settling time. However, the asymptotes [1 – exp (−ρt)] and [1 + exp (−ρt)] clearly impact settling time, and they are controlled by the time constants of the open-loop amplifier, particularly the shorter of the two time constants. That suggests that a specification on settling time must be met by appropriate design of the open-loop amplifier.

The two major conclusions from this analysis are:

1. Feedback controls the amplitude of oscillation about final value for a given open-loop amplifier and given values of open-loop time constants, τ_1 and τ_2.
2. The open-loop amplifier decides settling time. It sets the time scale of figure, and the faster the open-loop amplifier, the faster this time scale.

As an aside, it may be noted that real-world departures from this linear two-pole model occur due to two major complications: first, real amplifiers have more than two poles, as well as zeros; and second, real amplifiers are nonlinear, so their step response changes with signal amplitude.

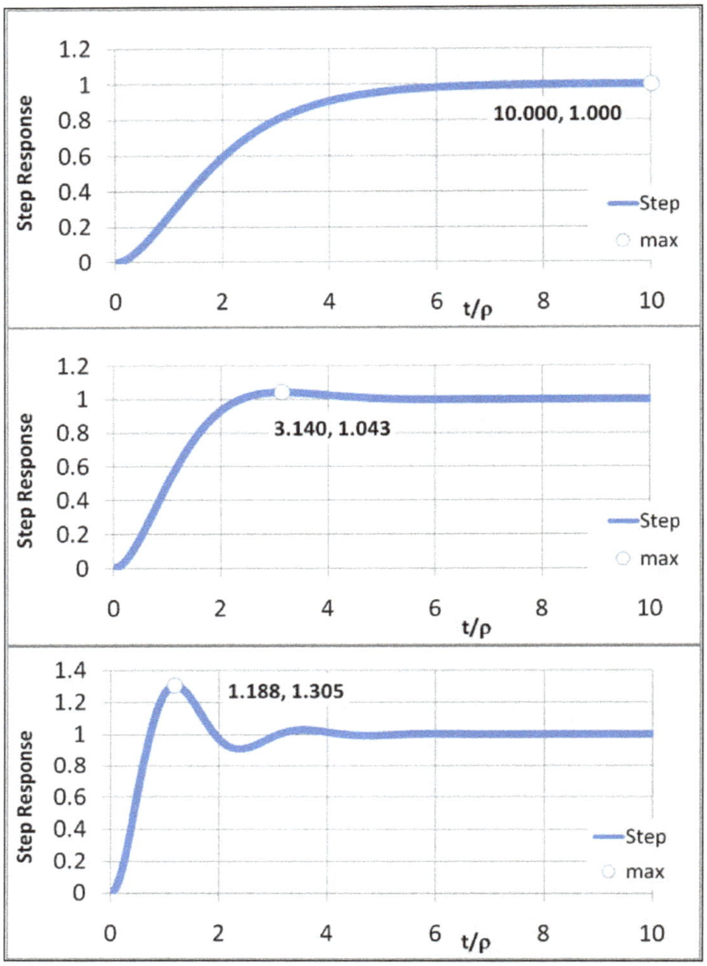

Step response for three values of α. Top: α = 4; Center: α = 2; Bottom: α = 0.5. As α is reduced the pole separation reduces, and the overshoot increases.

Control of Overshoot

Using the equations above, the amount of overshoot can be found by differentiating the step response and finding its maximum value. The result for maximum step response S_{max} is:

$$S_{max} = 1 + \exp\left(-\pi\frac{\rho}{\mu}\right).$$

The final value of the step response is 1, so the exponential is the actual overshoot itself. It is clear the overshoot is zero if $\mu = 0$, which is the condition:

$$\frac{4\beta A_0}{\tau_1\tau_2} = \left(\frac{1}{\tau_1} - \frac{1}{\tau_2}\right)^2.$$

This quadratic is solved for the ratio of time constants by setting $x = (\tau_1 / \tau_2)^{1/2}$ with the result,

$$x = \sqrt{\beta A_0} + \sqrt{\beta A_0 + 1}.$$

Because $\beta A_0 >> 1$, the 1 in the square root can be dropped, and the result is,

$$\frac{\tau_1}{\tau_2} = 4\beta A_0.$$

In words, the first time constant must be much larger than the second. To be more adventurous than a design allowing for no overshoot we can introduce a factor α in the above relation:

$$\frac{\tau_1}{\tau_2} = \alpha\beta A_0,$$

and let α be set by the amount of overshoot that is acceptable.

Figure illustrates the procedure. Comparing the top panel (α = 4) with the lower panel (α = 0.5) shows lower values for α increase the rate of response, but increase overshoot. The case α = 2 (center panel) is the *maximally flat* design that shows no peaking in the Bode gain vs. frequency plot. That design has the rule of thumb built-in safety margin to deal with non-ideal realities like multiple poles (or zeros), nonlinearity (signal amplitude dependence) and manufacturing variations, any of which can lead to too much overshoot. The adjustment of the pole separation (that is, setting α) is the subject of frequency compensation, and one such method is pole splitting.

Control of Settling Time

The amplitude of ringing in the step response in figure is governed by the damping factor exp ($-\rho t$). That is, if we specify some acceptable step response deviation from final value, say Δ, that is:

$$S(t) \leq 1 + \Delta$$

this condition is satisfied regardless of the value of βA_{OL} provided the time is longer than the settling time, say t_S, given by:

$$\Delta = e^{-\rho t_S} \text{ or } t_S = \frac{\ln\left(\frac{1}{\Delta}\right)}{\rho} = \tau_2 \frac{2\ln\left(\frac{1}{\Delta}\right)}{1 + \frac{\tau_2}{\tau_1}} \approx 2\tau_2 \ln\left(\frac{1}{\Delta}\right),$$

where the $\tau_1 >> \tau_2$ is applicable because of the overshoot control condition, which makes $\tau_1 = \alpha\beta A_{OL} \tau_2$. Often the settling time condition is referred to by saying the settling period is inversely proportional to the unity gain bandwidth, because $1/(2\pi \tau_2)$ is close to this bandwidth for an amplifier with typical dominant pole compensation. However, this result is more precise than this rule of thumb. As an example of this formula, if $\Delta = 1/e^4 = 1.8$ %, the settling time condition is $t_S = 8 \tau_2$.

In general, control of overshoot sets the time constant ratio, and settling time t_S sets τ_2.

System Identification using the Step Response: System with Two Real Poles

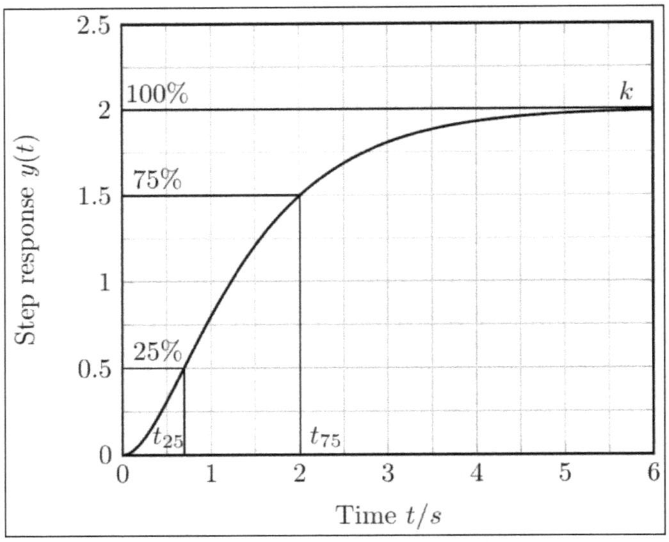

Step response of the system with . Measure the significant point, and .

This method uses significant points of the step response. There is no need to guess tangents to the measures Signal. The equations are derived using numerical simulations, determining some significant ratios and fitting parameters of nonlinear equations.

Here the steps:

- Measure the system step-response $y(t)$ of the system with an input step signal $x(t)$.
- Determine the time-spans t_{25} and t_{75} where the step response reaches 25% and 75% of the steady state output value.
- Determine the system steady-state gain $k = A_0$ with $k = \lim_{t\to\infty} \frac{y(t)}{x(t)}$
- Calculate

$$r = \frac{t_{25}}{t_{75}}$$

$$P = -18.56075r + \frac{0.57311}{r - 0.20747} + 4.16423$$

$$X = 14.2797r^3 - 9.3891r^2 + 0.25437r + 1.32148$$

- Determine the two time constants

$$\tau_2 = T_2 = \frac{t_{75} - t_{25}}{X(1 + 1/P)}$$

$$\tau_1 = T_1 = \frac{T_2}{P}$$

- Calculate the transfer function of the identified system within the Laplace-domain

$$G(s)=\frac{k}{(1+sT_1)\cdot(1+sT_2)}$$

Phase Margin

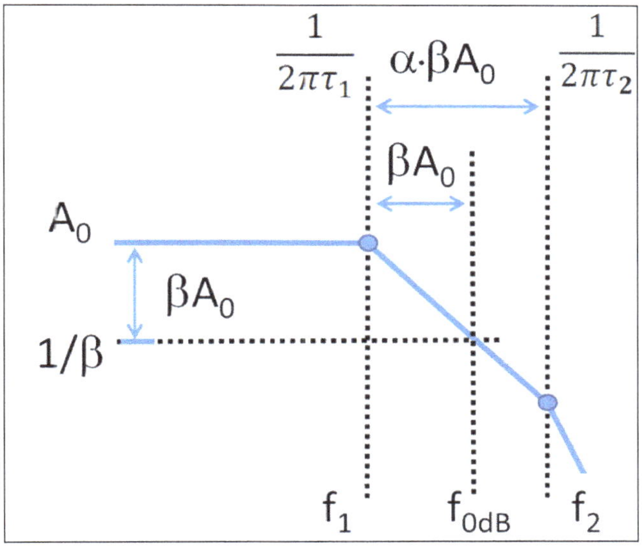

Bode gain plot to find phase margin; scales are logarithmic, so labeled separations are multiplicative factors. For example, $f_{0\,dB} = \beta A_0 \times f_1$.

Next, the choice of pole ratio τ_1/τ_2 is related to the phase margin of the feedback amplifier. Figure is the Bode gain plot for the two-pole amplifier in the range of frequencies up to the second pole position. The assumption behind figure is that the frequency $f_{0\,dB}$ lies between the lowest pole at $f_1 = 1/(2\pi\tau_1)$ and the second pole at $f_2 = 1/(2\pi\tau_2)$. As indicated in figure, this condition is satisfied for values of $\alpha \geq 1$.

Using figure the frequency (denoted by $f_{0\,dB}$) is found where the loop gain βA_0 satisfies the unity gain or 0 dB condition, as defined by:

$$|\beta A_{OL}(f_{0\,db})|=1.$$

The slope of the downward leg of the gain plot is (20 dB/decade); for every factor of ten increase in frequency, the gain drops by the same factor:

$$f_{0\,dB}=\beta A_0 f_1.$$

The phase margin is the departure of the phase at $f_{0\,dB}$ from −180°. Thus, the margin is:

$$\phi_m = 180^\circ - \arctan(f_{0\,dB}/f_1) - \arctan(f_{0\,dB}/f_2).$$

Because $f_{0\,dB}/f_1 = \beta A_0 >> 1$, the term in f_1 is 90°. That makes the phase margin:

$$\phi_m = 90^\circ - \arctan(f_{0\,dB}/f_2)$$

$$= 90^{\circ} - \arctan\left(\frac{\beta A_0 f_1}{\alpha \beta A_0 f_1}\right)$$

$$= 90^{\circ} - \arctan\left(\frac{1}{\alpha}\right) = \arctan(\alpha).$$

In particular, for case α = 1, φ_m = 45°, and for α = 2, φ_m = 63.4°. Sansen recommends α = 3, φ_m = 71.6° as a "good safety position to start with".

If α is increased by shortening τ_2, the settling time t_S also is shortened. If α is increased by lengthening τ_1, the settling time t_S is little altered. More commonly, both τ_1 *and* τ_2 change, for example if the technique of pole splitting is used.

As an aside, for an amplifier with more than two poles, the diagram of figure still may be made to fit the Bode plots by making f_2 a fitting parameter, referred to as an "equivalent second pole" position.

Formal Mathematical Description

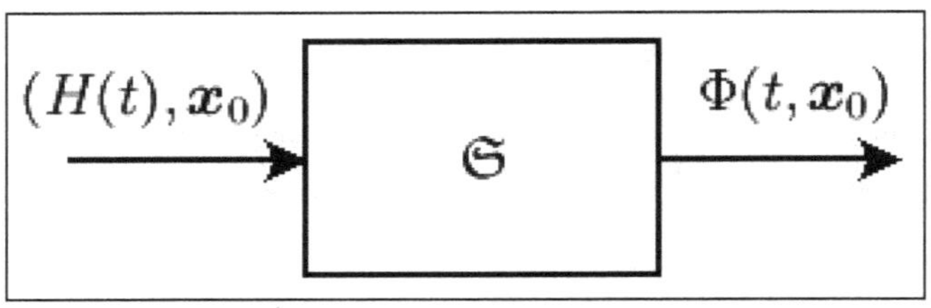

Black box representation of a dynamical system, its input and its step response.

This section provides a formal mathematical definition of step response in terms of the abstract mathematical concept of a dynamical system $\mathfrak{S}$: all notations and assumptions required for the following description are listed here.

- $t \in T$ is the evolution parameter of the system, called "time" for the sake of simplicity,
- $x|_t \in M$ is the state of the system at time t, called "output" for the sake of simplicity,
- $\Phi : T \times M \to M$ is the dynamical system evolution function,
- $\Phi(0, \mathbf{x}) = \mathbf{x}_0 \in M$ is the dynamical system initial state,
- $H(t)$ is the Heaviside step function

Nonlinear Dynamical System

For a general dynamical system, the step response is defined as follows:

$$\mathbf{x}|_t = \Phi_{\{H(t)\}}(t, \mathbf{x}_0).$$

It is the evolution function when the control inputs (or source term, or forcing inputs) are Heaviside functions: the notation emphasizes this concept showing *H*(*t*) as a subscript.

Linear Dynamical System

For a linear time-invariant black box, let $\mathfrak{S} \equiv S$ for notational convenience: the step response can be obtained by convolution of the Heaviside step function control and the impulse response $h(t)$ of the system itself

$$a(t) = h * H(t) = H * h(t) = \int_{-\infty}^{+\infty} h(\tau)H(t-\tau)d\tau = \int_{-\infty}^{t} h(\tau)d\tau.$$

which for an LTI system is equivalent to just integrating the latter. Conversely, for an LTI system, the derivative of the step response yields the impulse response:

$$h(t) = \frac{d}{dt}a(t)$$

However, these simple relations are not true for a non-linear or time-variant system.

BIBO Stability

In signal processing, specifically control theory, bounded-input, bounded-output (BIBO) stability is a form of stability for linear signals and systems that take inputs. If a system is BIBO stable, then the output will be bounded for every input to the system that is bounded.

A signal is bounded if there is a finite value $B > 0$ such that the signal magnitude never exceeds B, that is

$|y[n]| \leq B \quad \forall n \in \mathbb{Z}$ for discrete-time signals, or

$|y(t)| \leq B \quad \forall t \in \mathbb{R}$ for continuous-time signals.

Time-domain Condition for Linear Time-invariant Systems

Continuous-time Necessary and Sufficient Condition

For a continuous time linear time-invariant (LTI) system, the condition for BIBO stability is that the impulse response be absolutely integrable, i.e., its L^1 norm exists.

$$\int_{-\infty}^{\infty} |h(t)|\mathrm{d}t = \|h\|_1 < \infty$$

Discrete-time sufficient condition

For a discrete time LTI system, the condition for BIBO stability is that the impulse response be absolutely summable, i.e., its ℓ^1 norm exists.

$$\sum_{n=-\infty}^{\infty} |h[n]| = \|h\|_1 < \infty$$

Proof of Sufficiency

Given a discrete time LTI system with impulse response $h[n]$ the relationship between the input $x[n]$ and the output $y[n]$ is

$$y[n] = h[n] * x[n]$$

where * denotes convolution. Then it follows by the definition of convolution

$$y[n] = \sum_{k=-\infty}^{\infty} h[k]x[n-k]$$

Let $\| x \|_\infty$ be the maximum value of $| x[n] |$, i.e., the L_∞ -norm.

$$|y[n]| = \left| \sum_{k=-\infty}^{\infty} h[n-k]x[k] \right|$$

$$\leq \sum_{k=-\infty}^{\infty} |h[n-k]||x[k]| \text{ (by the triangle inequality)}$$

$$\leq \sum_{k=-\infty}^{\infty} |h[n-k]| \, \| x \|_\infty$$

$$= \| x \|_\infty \sum_{k=-\infty}^{\infty} |h[n-k]|$$

$$= \| x \|_\infty \sum_{k=-\infty}^{\infty} |h[k]|$$

If $h[n]$ is absolutely summable, then $\sum_{k=-\infty}^{\infty} |h[k]| = \| h \|_1 < \infty$ and

$$\| x \|_\infty \sum_{k=-\infty}^{\infty} | h[k] | = \| x \|_\infty \| h \|_1$$

So if $h[n]$ is absolutely summable and $| x[n] |$ is bounded, then $| y[n] |$ is bounded as well because $\| x \|_\infty \| h \|_1 < \infty$.

The proof for continuous-time follows the same arguments.

Frequency-domain Condition for Linear Time-invariant Systems

Continuous-time Signals

For a rational and continuous-time system, the condition for stability is that the region of convergence (ROC) of the Laplace transform includes the imaginary axis. When the system is causal, the ROC is the open region to the right of a vertical line whose abscissa is the real part of the "largest pole", or the pole that has the greatest real part of any pole in the system. The real part of the largest pole defining the ROC is called the abscissa of convergence. Therefore, all poles of the system must be in the strict left half of the s-plane for BIBO stability.

This stability condition can be derived from the above time-domain condition as follows:

$$\begin{aligned}\int_{-\infty}^{\infty} |h(t)|dt &= \int_{-\infty}^{\infty} |h(t)||e^{-j\omega t}|dt \\ &= \int_{-\infty}^{\infty} |h(t)(1\cdot e)^{-j\omega t}|dt \\ &= \int_{-\infty}^{\infty} |h(t)(e^{\sigma+j\omega})^{-t}|dt \\ &= \int_{-\infty}^{\infty} |h(t)e^{-st}|dt\end{aligned}$$

where $s = \sigma + j\omega$ and $\mathrm{Re}(s) = \sigma = 0$.

The region of convergence must therefore include the imaginary axis.

Discrete-time Signals

For a rational and discrete time system, the condition for stability is that the region of convergence (ROC) of the z-transform includes the unit circle. When the system is causal, the ROC is the open region outside a circle whose radius is the magnitude of the pole with largest magnitude. Therefore, all poles of the system must be inside the unit circle in the z-plane for BIBO stability.

This stability condition can be derived in a similar fashion to the continuous-time derivation:

$$\begin{aligned}\sum_{n=-\infty}^{\infty} |h[n]| &= \sum_{n=-\infty}^{\infty} |h[n]|\left|e^{-j\omega n}\right| \\ &= \sum_{n=-\infty}^{\infty} \left|h[n](1\cdot e)^{-j\omega n}\right| \\ &= \sum_{n=-\infty}^{\infty} \left|h[n](re^{j\omega})^{-n}\right| \\ &= \sum_{n=-\infty}^{\infty} \left|h[n]z^{-n}\right|\end{aligned}$$

where $z = re^{j\omega}$ and $r = |z| = 1$..

The region of convergence must therefore include the unit circle.

Convolution of Signals

Convolution of signals occurs in several contexts in signal processing. The input-output characteristic of linear time-invariant (LTI) systems, the most widely used type of system in signal processing, is described entirely in terms of the impulse response of the system. The impulse response is the output of the system due to an impulse input signal. Given the input signal, the output of an LTI system is the convolution of the input signal with the impulse response of the system. Hence, convolution plays a key role in relating the input and output signals of an LTI system.

Convolution also arises when we analyze the effect of multiplying two signals in the frequency domain. If we multiply two signals in time, the Fourier representation for the product is the convolution of the Fourier representations of the individual signals. This type of analysis occurs when the discrete Fourier transform is applied to truncated (finite length) signals, and in the design of finite impulse response (FIR) filters.

Graphical Convolution of Signals

If we define $y[n]$ as the convolution of signals $x[n]$ and $h[n]$, then we can write:

$$y[n]=\sum_{k=-\infty}^{\infty}h[k]x[n-k]$$

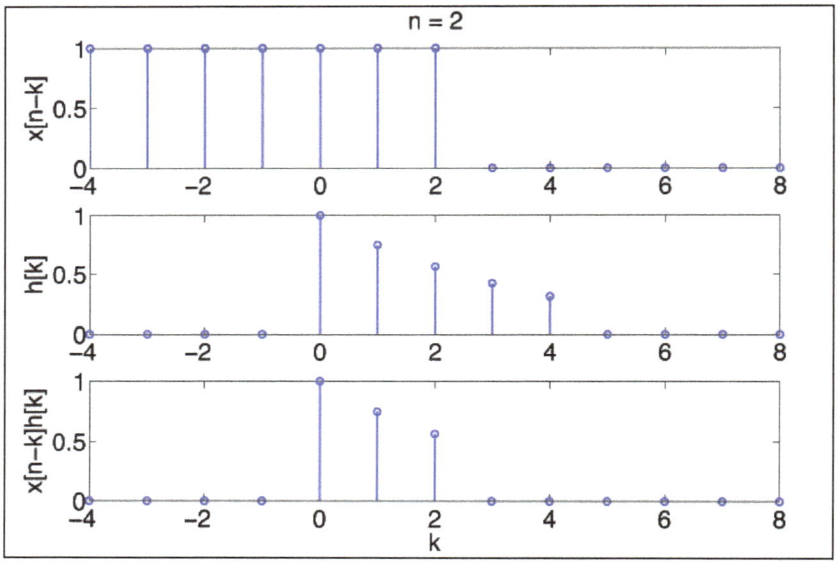

Forming h[k]x[n-k] for n=2 in an example.

In brief, first we form a signal $v_n[k]=h[k]x[n-k]$ for a desired value of n, and then we sum all the values in this signal over k to obtain $y[n]$. The graphical aspect of this procedure arises in the formation of $v_n[k]$. We graph $h[k]$ and graph $x[n-k]$, as a function of the independent variable k .Note that n is a constant in this step. In order to graph $x[n-k]$,first reflex $x[k]$ about $k=0$ to obtain $x[-k]$, then shift $x[-k]$ right by n values to obtain $x[n-k]$, It is usually quite simple to identify the product $v_n[k]=h[k]x[n-k]$ from the graphs of $h[k]$ and $x[n-k]$. This process is repeated for all values n where $y[n]$ is desired.

References

- Phillips, C.l., Parr, J.M., & Riskin, E.A (2007). Signals, systems and Transforms. Prentice Hall. ISBN 978-0-13-041207-2

- Digital-signal, communication-networks, computernetworkingnotes: ecomputernotes.com Plant-reproductive-system, science: britannica.com, Retrieved 23 June, 2019
- Mcclellan, James H.; Schafer, Ronald W.; Yoder, Mark A. (2015). DSP First, Second Edition. Pearson Education. P. 151. ISBN 978-0136019251
- Convolution-of-signals-why: allsignalprocessing.com, Retrieved 31 March, 2019
- F. Alton Everest (2000). Master Handbook of Acoustics (Fourth ed.). Mcgraw-Hill Professional. ISBN 0-07-136097-2

2

Understanding Digital Signal Processing

Digital signal processing deals with the diverse techniques which are used to improve the reliability and accuracy of digital connections. It uses computers and digital signal processors for processing the signals digitally. This chapter has been carefully written to provide an easy understanding of digital signal processing.

Digital signal processing (DSP) refers to various techniques for improving the accuracy and reliability of digital communications. The theory behind DSP is quite complex. Basically, DSP works by clarifying, or standardizing, the levels or states of a digital signal. ADSP circuit is able to differentiate between human-made signals,which are orderly, and noise, which is inherently chaotic.

All communications circuits contain some noise. This is true whether the signals are analog or digital,and regardless of the type of information conveyed. Noise is the eternal bane of communications engineers, who are always striving to find new ways to improve the signal-to-noise ratioin communications systems. Traditional methods of optimizing S/N ratio include increasing the transmitted signal power and increasing the receiver sensitivity. (In wireless systems,specialized antenna systems can also help.) Digital signal processing dramatically improves the sensitivity of a receiving unit.

The effect is most noticeable when noise competes with a desired signal. A good DSP circuit can sometimes seem like an electronic miracle worker. But there are limits to what it can do. If the noise is so strong that all traces of the signal are obliterated, a DSP circuit cannot find any order in the chaos, and no signal will be received.

If an incoming signal is analog, for example a standard television broadcast station, the signal is first converted to digital form by an analog-to-digital converter(ADC). The resulting digital signal has two or more levels. Ideally, these levels are always predictable, exact voltages or currents. However, because the incoming signal contains noise, the levels are not always at the standard values. The DSP circuit adjusts the levels so they are at the correct values. This practically eliminates the noise. The digital signal is then converted back to analog from via a digital-to-analog converter (DAC).

If a received signal is digital, for example computer data, then the ADC and DAC are not necessary. The DSP acts directly on the incoming signal, eliminating irregularities caused by noise, and thereby minimizing the number of errors per unit time.

Digital Signal Processor

A Digital Signal Processor, or DSP, is a semiconductor device used for processing signals digitally. A signal, in this context, traditionally refers to an analog signal (such as analog voltage) that has been converted into a digital one so that it can be processed mathematically. Nowadays, however, almost every piece of information has been digitized, so a digital signal may be any stream of digital data - digital audio/video data, betting odds, or even the weight of clothes in a washing machine. Analysis of such digital signals for a variety of purposes can be easily accomplished by a DSP.

Signal processing encompasses a large variety of actions performed on signals - filtering, encoding/decoding, compression/decompression, amplification, modulation, level detection, pattern matching, mathematical/logical operations, and much more. These processes are performed on a signal for a number of reasons: to enhance it; reduce its component noise; make its transmission and reception more effective, efficient, and faster; transform it; make it interact with other signals in special ways; facilitate its use in digital analysis, monitoring, or control; etc. A DSP has built-in capabilities to perform these signal processing functions easily.

A DSP is very similar to a microprocessor. In fact, it is regarded by many as a special microprocessor created particularly to process signals. Both a microprocessor and a DSP can execute instructions, accept input digital data, perform operations on them, and output digital data. The fundamental difference between a DSP and a microprocessor is what their built-in processing capabilities were designed for.

A DSP is a highly-specialized device that's equipped with a multitude of mathematical functions specifically intended for processing a digital signal, whereas a microprocessor is designed to be a general-purpose device. A microprocessor would be able to handle many different applications, such as word processing, spreadsheets, databases, and, well, even digital signal processing. However, it can not be as good as a DSP when it comes to serious DSP applications.

Current trends in technology seem to indicate the possibility though that the distinction between a DSP and a microprocessor will soon be gone. Microprocessors are becoming more and more sophisticated that some of them are now equipped with true DSP capabilities. It will just be a matter of time before high-end microprocessors will have the capability to perform high-end signal processing, or any high-end task for that matter.

A DSP is also very similar to a microprocessor as far as architecture is concerned, i.e., it has many parts that are also seen in a microprocessor, such as data and address buses, an Arithmetic-Logic Unit (ALU), a program control unit, assorted flags and registers, etc. It also has its own native instruction set, which defines what it can be programmed to do. Programming DSP's is no longer complicated too, with the existence of various development kits in the market that support DSP software development using high-level programming languages such as C.

Many DSP applications deal with real-world analog signals (such as sound, light, analog voltage, analog current, temperature, pressure). Since a DSP can only process digital signals, there is a need to convert analog signals first into digital data before they can be processed by a DSP. After processing, there is again a need for the DSP to convert these digital data back into the original real-world analog signal format. In such applications, the DSP must be supported by an analog-to-digital converter (ADC) and a digital-to-analog converter (DAC), which will perform the required analog-digital and digital-analog conversions, respectively.

Applications where DSP's are commonly used include: 1) digital sound and image processing; 2) digital communications; 3) consumer electronics (e.g., mobile phones, faxes, computer peripherals such as modems and sound cards, and digital entertainment systems such as DVD players and digital TV); 4) medical electronics; and 5) industrial and automation electronics.

Digital Signal Processor Architecture

One of the biggest bottlenecks in executing DSP algorithms is transferring information to and from memory. This includes data, such as samples from the input signal and the filter coefficients, as well as program instructions, the binary codes that go into the program sequencer. For example, suppose we need to multiply two numbers that reside somewhere in memory. To do this, we must fetch three binary values from memory, the numbers to be multiplied, plus the program instruction describing what to do.

This seemingly simple task is done in a traditional microprocessor. This is often called a Von Neumann architecture, after the brilliant American mathematician John Von Neumann. Von Neumann guided the mathematics of many important discoveries of the early twentieth century. His many achievements include: developing the concept of a stored program computer, formalizing the mathematics of quantum mechanics, and work on the atomic bomb. If it was new and exciting, Von Neumann was there!

As shown in below (a), a Von Neumann architecture contains a single memory and a single bus for transferring data into and out of the central processing unit (CPU). Multiplying two numbers requires at least three clock cycles, one to transfer each of the three numbers over the bus from the memory to the CPU. We don't count the time to transfer the result back to memory, because we assume that it remains in the CPU for additional manipulation (such as the sum of products in an FIR filter). The Von Neumann design is quite satisfactory when you are content to execute all of the required tasks in serial. In fact, most computers today are of the Von Neumann design. We only need other architectures when very fast processing is required, and we are willing to pay the price of increased complexity.

This leads us to the Harvard architecture, shown in below (b). This is named for the work done at Harvard University in the 1940s under the leadership of Howard Aiken. As shown in this illustration, Aiken insisted on separate memories for data and program instructions, with separate buses for each. Since the buses operate independently, program instructions and data can be fetched at the same time, improving the speed over the single bus design. Most present day DSPs use this dual bus architecture.

Figure illustrates the next level of sophistication, the Super Harvard Architecture. This term was coined by Analog Devices to describe the internal operation of their ADSP-2106x and new ADSP-211xx families of Digital Signal Processors. These are called SHARC DSPs, a contraction of the longer term, Super Harvard ARChitecture. The idea is to build upon the Harvard architecture by adding features to improve the throughput. While the SHARC DSPs are optimized in dozens of ways, two areas are important enough to be included in figure: an instruction cache, and an I/O controller.

First, let's look at how the instruction cache improves the performance of the Harvard architecture. A handicap of the basic Harvard design is that the data memory bus is busier than the program memory bus. When two numbers are multiplied, two binary values (the numbers) must be passed over the data memory bus, while only one binary value (the program instruction) is passed over the program memory bus. To improve upon this situation, we start by relocating part of the "data" to program memory. For instance, we might place the filter coefficients in program memory, while keeping the input signal in data memory. This relocated data is called "secondary data". At first glance, this doesn't seem to help the situation; now we must transfer one value over the data memory bus (the input signal sample), but two values over the program memory bus (the program instruction and the coefficient). In fact, if we were executing random instructions, this situation would be no better at all.

However, DSP algorithms generally spend most of their execution time in loops, such as instructions 6-12 of table. This means that the same set of program instructions will continually pass from program memory to the CPU. The Super Harvard architecture takes advantage of this situation by including an instruction cache in the CPU. This is a small memory that contains about 32 of the most recent program instructions. The first time through a loop, the program instructions must be passed over the program memory bus. This results in slower operation because of the conflict with the coefficients that must also be fetched along this path. However, on additional executions of the loop, the program instructions can be pulled from the instruction cache. This means that all of the memory to CPU information transfers can be accomplished in a single cycle: the sample from the input signal comes over the data memory bus, the coefficient comes over the program memory bus, and the program instruction comes from the instruction cache. In the jargon of the field, this efficient transfer of data is called a high memory-access bandwidth.

Figure presents a more detailed view of the SHARC architecture, showing the I/O controller connected to data memory. This is how the signals enter and exit the system. For instance, the SHARC DSPs provides both serial and parallel communications ports. These are extremely high speed connections. For example, at a 40 MHz clock speed, there are two serial ports that operate at 40 Mbits/second each, while six parallel ports each provide a 40 Mbytes/second data transfer. When all six parallel ports are used together, the data transfer rate is an incredible 240 Mbytes/ second.

This is fast enough to transfer the entire text of this book in only 2 milliseconds! Just as important, dedicated hardware allows these data streams to be transferred directly into memory (Direct Memory Access, or DMA), without having to pass through the CPU's registers. In other words, tasks 1 & 14 on our list happen independently and simultaneously with the other tasks; no cycles are stolen from the CPU. The main buses (program memory bus and data memory bus) are also accessible from outside the chip, providing an additional interface to off-chip memory and peripherals. This allows the SHARC DSPs to use a four Gigaword (16 Gbyte) memory, accessible at 40 Mwords/second (160 Mbytes/second), for 32 bit data.

This type of high speed I/O is a key characteristic of DSPs. The overriding goal is to move the data in, perform the math, and move the data out before the next sample is available. Everything else is secondary. Some DSPs have on-board analog-to-digital and digital-to-analog converters, a feature called mixed signal. However, all DSPs can interface with external converters through serial or parallel ports.

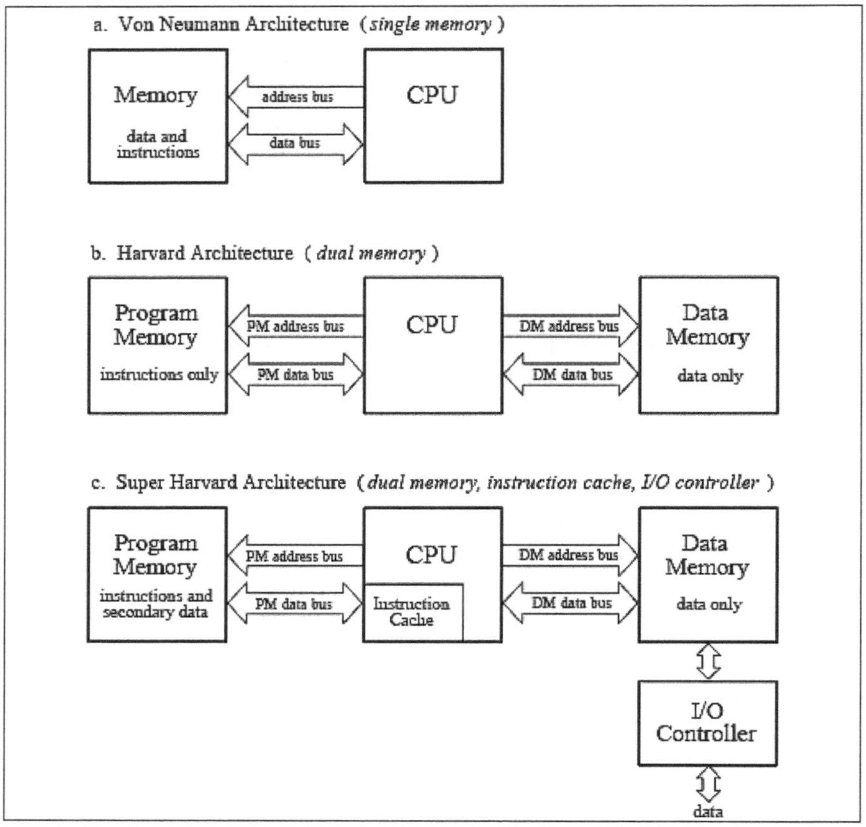

Microprocessor architecture, The Von Neumann architecture uses a single memory to hold both data and instructions. In comparison, the Harvard architecture uses separate memories for data and instructions, providing higher speed. The Super Harvard Architecture improves upon the Harvard deign by adding an instruction cache and a dedicated I/O controller.

Now let's look inside the CPU. At the top of the diagram are two blocks labeled Data Address Generator (DAG), one for each of the two memories. These control the addresses sent to the program and data memories, specifying where the information is to be read from or written to. In simpler microprocessors this task is handled as an inherent part of the program sequencer, and is quite transparent to the programmer. However, DSPs are designed to operate with circular buffers, and benefit from the extra hardware to manage them efficiently. This avoids needing to use precious CPU clock cycles to keep track of how the data are stored. For instance, in the SHARC DSPs, each of the two DAGs can control eight circular buffers. This means that each DAG holds 32 variables (4 per buffer), plus the required logic.

Why so many circular buffers? Some DSP algorithms are best carried out in stages. For instance, IIR filters are more stable if implemented as a cascade of biquads (a stage containing two poles and up to two zeros). Multiple stages require multiple circular buffers for the fastest operation. The DAGs in the SHARC DSPs are also designed to efficiently carry out the Fast Fourier transform. In this mode, the DAGs are configured to generate bit-reversed addresses into the circular buffers, a necessary part of the FFT algorithm. In addition, an abundance of circular buffers greatly simplifies DSP code generation- both for the human programmer as well as high-level language compilers, such as C.

The data register section of the CPU is used in the same way as in traditional microprocessors. In the ADSP-2106x SHARC DSPs, there are 16 general purpose registers of 40 bits each. These can hold intermediate calculations, prepare data for the math processor, serve as a buffer for data transfer, hold flags for program control, and so on. If needed, these registers can also be used to control loops and counters; however, the SHARC DSPs have extra hardware registers to carry out many of these functions.

The math processing is broken into three sections, a multiplier, an arithmetic logic unit (ALU), and a barrel shifter. The multiplier takes the values from two registers, multiplies them, and places the result into another register. The ALU performs addition, subtraction, absolute value, logical operations (AND, OR, XOR, NOT), conversion between fixed and floating point formats, and similar functions. Elementary binary operations are carried out by the barrel shifter, such as shifting, rotating, extracting and depositing segments, and so on. A powerful feature of the SHARC family is that the multiplier and the ALU can be accessed in parallel. In a single clock cycle, data from registers 0-7 can be passed to the multiplier, data from registers 8-15 can be passed to the ALU, and the two results returned to any of the 16 registers.

There are also many important features of the SHARC family architecture that aren't shown in this simplified illustration. For instance, an 80 bit accumulator is built into the multiplier to reduce the round-off error associated with multiple fixed-point math operations. Another interesting feature is the use of shadow registers for all the CPU's key registers. These are duplicate registers that can be switched with their counterparts in a single clock cycle. They are used for fast context switching, the ability to handle interrupts quickly. When an interrupt occurs in traditional microprocessors, all the internal data must be saved before the interrupt can be handled. This usually involves pushing all of the occupied registers onto the stack, one at a time. In comparison, an interrupt in the SHARC family is handled by moving the internal data into the shadow registers in a single clock cycle. When the interrupt routine is completed, the registers are just as quickly restored. This feature allows step 4 on our list (managing the sample-ready interrupt) to be handled very quickly and efficiently.

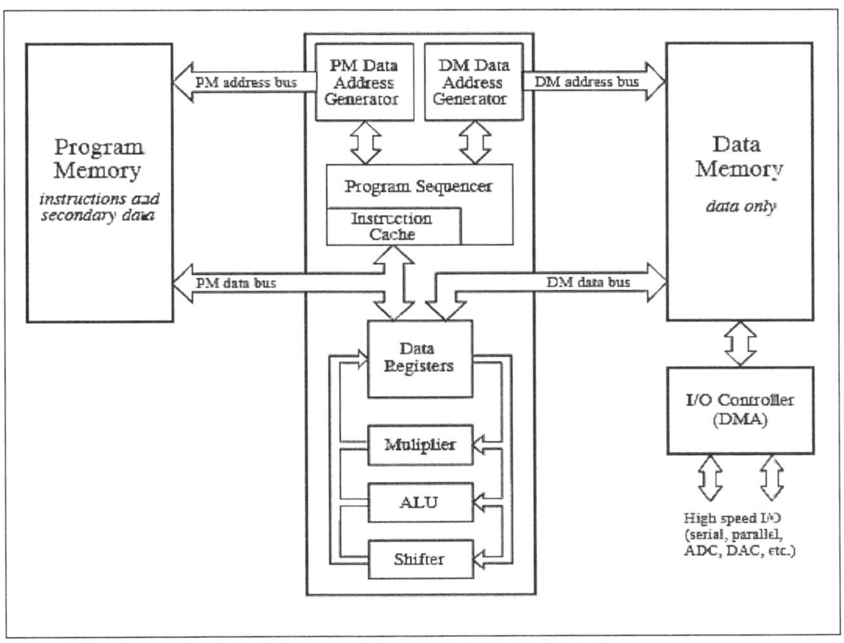

Typical DSP architecture. Digital Signal Processors are designed to implement tasks in parallel. This simplified diagram is of the Analog Devices SHARC DSP. Compare this architecture with the tasks needed to implement an FIR filter, as listed in Table. All of the steps within the loop can be executed In a single clock cycle.

Now we come to the critical performance of the architecture, how many of the operations within the loop can be carried out at the same time. Because of its highly parallel nature, the SHARC DSP can simultaneously carry out all of these tasks. Specifically, within a single clock cycle, it can perform a multiply, an addition, two data moves, update two circular buffer pointers (steps 8 and 10), and control the loop. There will be extra clock cycles associated with beginning and ending the loop; however, these tasks are also handled very efficiently. If the loop is executed more than a few times, this overhead will be negligible. As an example, suppose you write an efficient FIR filter program using 100 coefficients. You can expect it to require about 105 to 110 clock cycles per sample to execute (i.e., 100 coefficient loops plus overhead). This is very impressive; a traditional microprocessor requires many thousands of clock cycles for this algorithm.

3

Signal Sampling

The process of mapping input values from a large set to output values in a smaller set is referred to as signal sampling quantization. Analog-to-digital conversion and digital to analog conversion are the two main types of digital conversions. The topics elaborated in this chapter will help in gaining a better perspective about these types and related aspects of signal sampling quantization.

A continuous signal $x(t)$ can be sampled by an analog-to-digital converter (A/D) to become a discrete signal composed of a set of time samples $x[m] = x(mt_0) = x(m/F)$ $(m = \cdots, -1, 0, 1, \cdots)$, where t_0 is the sampling period, the time interval between two consecutive samples, and $F = 1/t_0$ is the sampling frequency.

The sampled signal $x_s(t)$ can be represented as the product of the time signal and an impulse train, the sampling function (also called a comb function):

$$x_s(t) = x(t) \cdot comb(t) = x(t) \cdot \sum_{m=-\infty}^{\infty} \delta(t - mt_0) = \sum_{m=-\infty}^{\infty} x[m]\delta(t - mt_0)$$

The Fourier transform of this sampled signal can be found as:

$$X_s(j\omega) = \int_{-\infty}^{\infty} x_s(t) e^{-j\omega t} dt = \int_{-\infty}^{\infty} [\sum_{m=-\infty}^{\infty} x[m]\delta(t - mt_o)] e^{-j\omega t} dt$$

$$= \sum_{m=-\infty}^{\infty} x[m] \int_{-\infty}^{\infty} \delta(t - mt_o) e^{-j\omega t} dt = \sum_{m=-\infty}^{\infty} x[m] e^{-j\omega m t_o} = \sum_{m=-\infty}^{\infty} x[m] e^{-j2\pi f m t_o}$$

$$= X_s(\omega) = X_s(f)$$

Note that this spectrum is periodic with period $F = 1/t_0$:

$$X_s(f + F) = \sum_{m=-\infty}^{\infty} x[m] e^{-j2\pi f m t_0} e^{-j2\pi F m t_0} = \sum_{m=-\infty}^{\infty} x[m] e^{-j2\pi f m t_0} = X_s(f)$$

According to convolution theorem, the sampling process as a multiplication in time domain becomes a convolution in frequency domain. Let,

$$X(f) = F[x(t)], \quad \text{and} \quad Comb(f) = F[comb(t)] = F \sum_{n=-\infty}^{\infty} \delta(f - nF)$$

then the spectrum of the sampled signal is:

$$X_s(f) = X(f) * F \sum_{n=-\infty}^{\infty} \delta(f - nF)$$

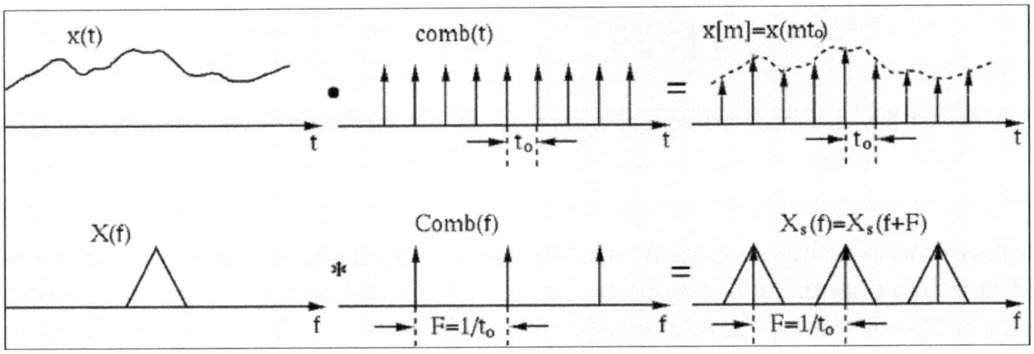

If we assume unit sampling rate $F = 1$ and therefore $t_0 = 1/F = 1$, then,

$$X_s(f) = \sum_{m=-\infty}^{\infty} x[m]e^{-j2\pi fm}, \quad \text{or} \quad X_s(\omega) = \sum_{m=-\infty}^{\infty} x[m]e^{-j\omega m}$$

Time samples of the discrete signal can be obtained from the spectrum $X_s(\omega)$. We multiply $\underline{e^{j\omega n}}$ on both sides of the above equation and integrate with respect to f over $\underline{0}$ to $\underline{1}$:

$$\int_0^1 X_s(\omega)e^{j\omega n}df = \sum_{m=-\infty}^{\infty} x[m]\int_0^1 e^{-j\omega(m-n)f}df = \sum_{m=-\infty}^{\infty} x[m]\delta(m-n) = x[n]$$

due to,

$$\int_0^1 e^{-j\omega kt}df = \int_0^1 e^{-j2k\pi f}df = \int_0^1 e^{-j2k\pi f}dt = \int_0^1 \cos(2\pi kt)dt + j\int_0^1 \sin(2\pi kt)dt = \delta[k]$$

(Integral of these sinusoids of period $1/k$ over $\underline{1}$ is zero except when $\underline{k=0}$ when $\cos(2\pi 0t) = 1$.)

The equation above is the inverse Fourier transform. We rewrite these two equations below:

$$X_s(\omega) = \sum_{n=-\infty}^{\infty} x[n]e^{-j\omega n} \sum_{n=-\infty}^{\infty} x[n]e^{-j2\pi fn} = X_s(f)$$

$$x[n] = \int_0^1 X_s(f)e^{-j2\pi fn}df = \frac{1}{2\pi}\int_0^{2\pi} X_s(\omega)e^{j\omega n}d\omega$$

This is the discrete-time Fourier transform (DTFT), which is sometimes also expressed as a function of $e^{j\omega} = e^{j2\pi f}$:

$$X(e^{j\omega}) = \sum_{n=-\infty}^{\infty} x[n]e^{-jn\omega}$$

$$x[n] = \frac{1}{2\pi}\int_0^{2\pi} X(e^{j\omega})e^{jn\omega}d\omega$$

Nyquist–Shannon Sampling Theorem

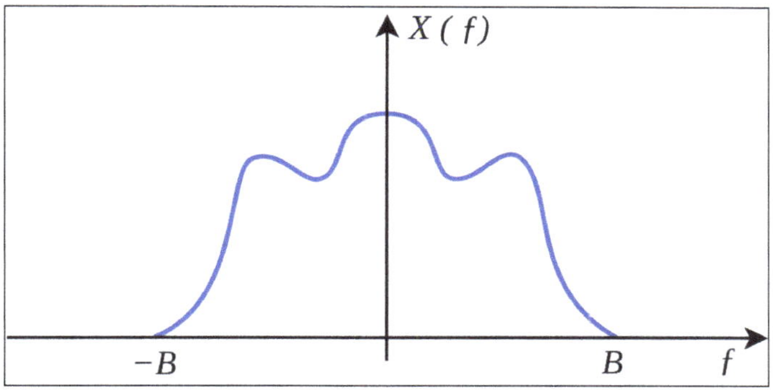

Example of magnitude of the Fourier transform of a bandlimited function.

In the field of digital signal processing, the sampling theorem is a fundamental bridge between continuous-time signals and discrete-time signals. It establishes a sufficient condition for a sample rate that permits a discrete sequence of *samples* to capture all the information from a continuous-time signal of finite bandwidth.

Strictly speaking, the theorem only applies to a class of mathematical functions having a Fourier transform that is zero outside of a finite region of frequencies. Intuitively we expect that when one reduces a continuous function to a discrete sequence and interpolates back to a continuous function, the fidelity of the result depends on the density (or sample rate) of the original samples. The sampling theorem introduces the concept of a sample rate that is sufficient for perfect fidelity for the class of functions that are bandlimited to a given bandwidth, such that no actual information is lost in the sampling process. It expresses the sufficient sample rate in terms of the bandwidth for the class of functions. The theorem also leads to a formula for perfectly reconstructing the original continuous-time function from the samples.

Perfect reconstruction may still be possible when the sample-rate criterion is not satisfied, provided other constraints on the signal are known. In some cases (when the sample-rate criterion is not satisfied), utilizing additional constraints allows for approximate reconstructions. The fidelity of these reconstructions can be verified and quantified utilizing Bochner's theorem.

The name *Nyquist–Shannon sampling theorem* honours Harry Nyquist and Claude Shannon albeit the fact that it had already been discovered in 1933 by Vladimir Kotelnikov. The theorem was also discovered independently by E. T. Whittaker and by others. It is thus also known by the names *Nyquist–Shannon–Kotelnikov, Whittaker–Shannon–Kotelnikov, Whittaker–Nyquist–Kotelnikov–Shannon*, and *cardinal theorem of interpolation.*

Sampling is a process of converting a signal (for example, a function of continuous time and space) into a sequence of values (a function of discrete time and space). Shannon's version of the theorem states:

> If a function $x(t)$ contains no frequencies higher than B hertz, it is completely determined by giving its ordinates at a series of points spaced $1/(2B)$ seconds apart.

A sufficient sample-rate is therefore anything larger than samples per second. Equivalently, $2B$ for a given sample rate f_s, perfect reconstruction is guaranteed possible for a bandlimit $B < f_s / 2$.

When the bandlimit is too high (or there is no bandlimit), the reconstruction exhibits imperfections known as aliasing. Modern statements of the theorem are sometimes careful to explicitly state that $x(t)$ must contain no sinusoidal component at exactly frequency *B*, or that *B* must be strictly less than ½ the sample rate. The threshold $2B$ is called the Nyquist rate and is an attribute of the continuous-time input $x(t)$ to be sampled. The sample rate must exceed the Nyquist rate for the samples to suffice to represent *x*(*t*). The threshold $f_s/2$ is called the Nyquist frequency and is an attribute of the sampling equipment. All meaningful frequency components of the properly sampled *x*(*t*) exist below the Nyquist frequency. The condition described by these inequalities is called the Nyquist criterion, or sometimes the *Raabe condition.* The theorem is also applicable to functions of other domains, such as *space,* in the case of a digitized image. The only change, in the case of other domains, is the units of measure applied to t, f_s, and *B*.

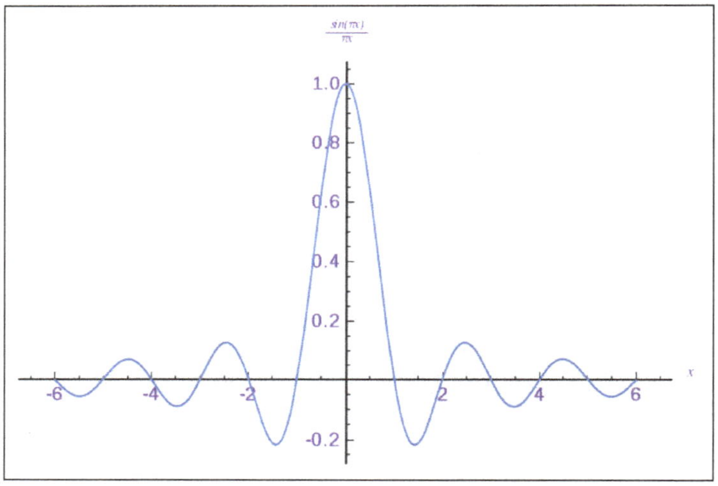

The normalized sinc function: $\sin(\pi x) / (\pi x)$ showing the central peak at $x = 0$, and zero-crossings at the other integer values of *x*.

The symbol $T = 1/f_s$ is customarily used to represent the interval between samples and is called the sample period or sampling interval. And the samples of function *x*(*t*) are commonly denoted by $x[n] = x(nT)$ (alternatively "x_n" in older signal processing literature), for all integer values of *n*. A mathematically ideal way to interpolate the sequence involves the use of sinc functions. Each sample in the sequence is replaced by a sinc function, centered on the time axis at the original location of the sample, *nT*, with the amplitude of the sinc function scaled to the sample value, *x*[*n*]. Subsequently, the sinc functions are summed into a continuous function. A mathematically equivalent method is to convolve one sinc function with a series of Dirac delta pulses, weighted by the sample values. Neither method is numerically practical. Instead, some type of approximation of the sinc functions, finite in length, is used. The imperfections attributable to the approximation are known as interpolation error.

Practical digital-to-analog converters produce neither scaled and delayed sinc functions, nor ideal Dirac pulses. Instead they produce a piecewise-constant sequence of scaled and delayed rectangular pulses (the zero-order hold), usually followed by an "anti-aliasing filter" to clean up spurious high-frequency content.

Derivation as a Special Case of Poisson Summation

When there is no overlap of the copies (also known as "images") of $X(f)$, the $k=0$ term of,

$$X_s(f) \triangleq \sum_{k=-\infty}^{\infty} X(f-kf_s) = \sum_{n=-\infty}^{\infty} T \cdot x(nT)\, e^{-i2\pi nTf}, X_s(f) \triangleq \sum_{k=-\infty}^{\infty} X(f-kf_s) = \sum_{n=-\infty}^{\infty} T \cdot x(nT)\, e^{-i2\pi nTf},$$

can be recovered by the product:

$$X(f) = H(f) \cdot X_s(f),$$

where:

$$H(f) \triangleq \begin{cases} 1 & |f| < B \\ 0 & |f| > f_s - B. \end{cases}$$

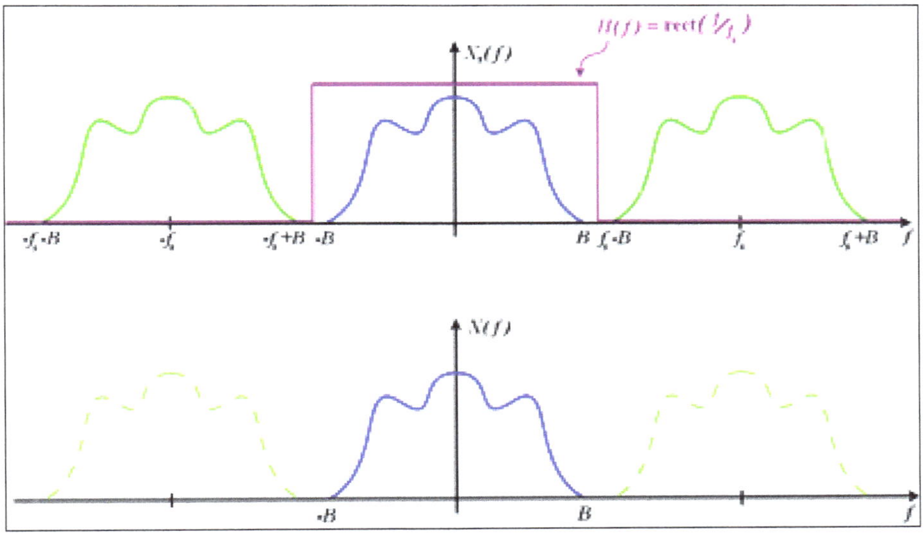

Spectrum, $X_s(f)$, of a properly sampled bandlimited signal (blue) and the adjacent DTFT images (green) that do not overlap. A brick-wall low-pass filter, $H(f)$, removes the images, leaves the original spectrum, $X(f)$, and recovers the original signal from its samples.

The sampling theorem is proved since $X(f)$ uniquely determines $x(t)$.

All that remains is to derive the formula for reconstruction. $H(f)$ need not be precisely defined in the region $[B, f_s - B]$ because $X_s(f)$ is zero in that region. However, the worst case is when $B = f_s/2$, the Nyquist frequency. A function that is sufficient for that and all less severe cases is:

$$H(f) = \mathrm{rect}\left(\frac{f}{f_s}\right) = \begin{cases} 1 & |f| < \frac{f_s}{2} \\ 0 & |f| > \frac{f_s}{2}, \end{cases}$$

where rect(•) is the rectangular function. Therefore:

$$X(f) = \mathrm{rect}\left(\frac{f}{f_s}\right) \cdot X_s(f)$$

$$= \text{rect}(Tf) \cdot \sum_{n=-\infty}^{\infty} T \cdot x(nT)\, e^{-i2\pi nTf}$$

$$= \sum_{n=-\infty}^{\infty} x(nT) \cdot \underbrace{T \cdot \text{rect}(Tf) \cdot e^{-i2\pi nTf}}_{\mathcal{F}\left\{\text{sinc}\left(\frac{t-nT}{T}\right)\right\}}$$

The inverse transform of both sides produces the Whittaker–Shannon interpolation formula:

$$x(t) = \sum_{n=-\infty}^{\infty} x(nT) \cdot \text{sinc}\left(\frac{t-nT}{T}\right),$$

which shows how the samples, $x(nT)$, can be combined to reconstruct $x(t)$.

- Larger-than-necessary values of f_s (smaller values of T), called *oversampling*, have no effect on the outcome of the reconstruction and have the benefit of leaving room for a *transition band* in which $H(f)$ is free to take intermediate values. Undersampling, which causes aliasing, is not in general a reversible operation.
- Theoretically, the interpolation formula can be implemented as a low pass filter, whose impulse response is sinc(t/T) and whose input is $\sum_{n=-\infty}^{\infty} x(nT) \cdot \delta(t-nT)$, which is a Dirac comb function modulated by the signal samples. Practical digital-to-analog converters (DAC) implement an approximation like the zero-order hold. In that case, oversampling can reduce the approximation error.

Shannon's Original Proof

Poisson shows that the Fourier series in produces the periodic summation of $X(f)$, regardless of f_s and B. Shannon, however, only derives the series coefficients for the case $f_s = 2B$. Virtually quoting Shannon's original paper:

Let $X(\omega)$ be the spectrum of $x(t)$. Then,

$$x(t) = \frac{1}{2\pi}\int_{-\infty}^{\infty} X(\omega) e^{i\omega t}\, d\omega = \frac{1}{2\pi}\int_{-2\pi B}^{2\pi B} X(\omega) e^{i\omega t}\, d\omega,$$

because $X(\omega)$ is assumed to be zero outside the band $\left|\frac{\omega}{2\pi}\right| < B$. If we let $t = \frac{n}{2B}$, where n is any positive or negative integer, we obtain:

$$x\left(\frac{n}{2B}\right) = \frac{1}{2\pi}\int_{-2\pi B}^{2\pi B} X(\omega) e^{i\omega \frac{n}{2B}}\, d\omega.$$

On the left are values of $x(t)$ at the sampling points. The integral on the right will be recognized as essentially the nth coefficient in a Fourier-series expansion of the function $X(\omega)$, taking the interval $-B$ to B as a fundamental period. This means that the values of the samples $x(n/2B)$ determine the Fourier coefficients in the series expansion of $X(\omega)$ Thus they determine $X(\omega)$ since $X(\omega)$ is zero for frequencies greater than B, and for lower frequencies $X(\omega)$ is determined if its Fourier coefficients are determined. But $X(\omega)$ determines the original function $x(t)$ completely, since a function is determined if its spectrum is known. Therefore the original samples determine the function $x(t)$ completely.

Shannon's proof of the theorem is complete at that point, but he goes on to discuss reconstruction via sinc functions, what we now call the Whittaker–Shannon interpolation formula. He does not derive or prove the properties of the sinc function, but these would have been familiar to engineers reading his works at the time, since the Fourier pair relationship between rect (the rectangular function) and sinc was well known.

Let x_n be the nth sample. Then the function $x(t)$ is represented by:

$$x(t) = \sum_{n=-\infty}^{\infty} x_n \frac{\sin \pi(2Bt - n)}{\pi(2Bt - n)}.$$

As in the other proof, the existence of the Fourier transform of the original signal is assumed, so the proof does not say whether the sampling theorem extends to bandlimited stationary random processes.

Application to Multivariable Signals and Images

Subsampled image showing a Moiré pattern.

Properly sampled image.

The sampling theorem is usually formulated for functions of a single variable. Consequently, the theorem is directly applicable to time-dependent signals and is normally formulated in that context. However, the sampling theorem can be extended in a straightforward way to functions of arbitrarily many variables. Grayscale images, for example, are often represented as two-dimensional arrays (or matrices) of real numbers representing the relative intensities of pixels (picture elements) located at the intersections of row and column sample locations. As a result, images require two independent variables, or indices, to specify each pixel uniquely—one for the row, and one for the column.

Color images typically consist of a composite of three separate grayscale images, one to represent each of the three primary colors—red, green, and blue, or *RGB* for short. Other colorspaces using 3-vectors for colors include HSV, CIELAB, XYZ, etc. Some colorspaces such as cyan, magenta, yellow, and black (CMYK) may represent color by four dimensions. All of these are treated as vector-valued functions over a two-dimensional sampled domain.

Similar to one-dimensional discrete-time signals, images can also suffer from aliasing if the sampling resolution, or pixel density, is inadequate. For example, a digital photograph of a striped shirt with high frequencies (in other words, the distance between the stripes is small), can cause aliasing of the shirt when it is sampled by the camera's image sensor. The aliasing appears as a moiré pattern. The "solution" to higher sampling in the spatial domain for this case would be to move closer to the shirt, use a higher resolution sensor, or to optically blur the image before acquiring it with the sensor.

Another example is shown to the right in the brick patterns. The top image shows the effects when the sampling theorem's condition is not satisfied. When software rescales an image it, in effect, runs the image through a low-pass filter first and then downsamples the image to result in a smaller image that does not exhibit the moiré pattern. The top image is what happens when the image is downsampled without low-pass filtering.

The sampling theorem applies to camera systems, where the scene and lens constitute an analog spatial signal source, and the image sensor is a spatial sampling device. Each of these components is characterized by a modulation transfer function (MTF), representing the precise resolution (spatial bandwidth) available in that component. Effects of aliasing or blurring can occur when the lens MTF and sensor MTF are mismatched. When the optical image which is sampled by the sensor device contains higher spatial frequencies than the sensor, the under sampling acts as a low-pass filter to reduce or eliminate aliasing. When the area of the sampling spot (the size of the pixel sensor) is not large enough to provide sufficient spatial anti-aliasing, a separate anti-aliasing filter (optical low-pass filter) may be included in a camera system to reduce the MTF of the optical image. Instead of requiring an optical filter, the graphics processing unit of smartphone cameras performs digital signal processing to remove aliasing with a digital filter. Digital filters also apply sharpening to amplify the contrast from the lens at high spatial frequencies, which otherwise falls off rapidly at diffraction limits.

The sampling theorem also applies to post-processing digital images, such as to up or down sampling. Effects of aliasing, blurring, and sharpening may be adjusted with digital filtering implemented in software, which necessarily follows the theoretical principles.

Critical Frequency

To illustrate the necessity of $f_s > 2B$,, consider the family of sinusoids generated by different values of θ in this formula:

$$x(t) = \frac{\cos(2\pi Bt + \theta)}{\cos(\theta)} = \cos(2\pi Bt) - \sin(2\pi Bt)\tan(\theta), \quad -\pi/2 < \theta < \pi/2.$$

With $f_s = 2B$ or equivalently $T = 1/2B$, the samples are given by:

$$x(nT) = \cos(\pi n) - \underbrace{\sin(\pi n)}_{0}\tan(\theta) = (-1)^n$$

regardless of the value of θ. That sort of ambiguity is the reason for the *strict* inequality of the sampling theorem's condition.

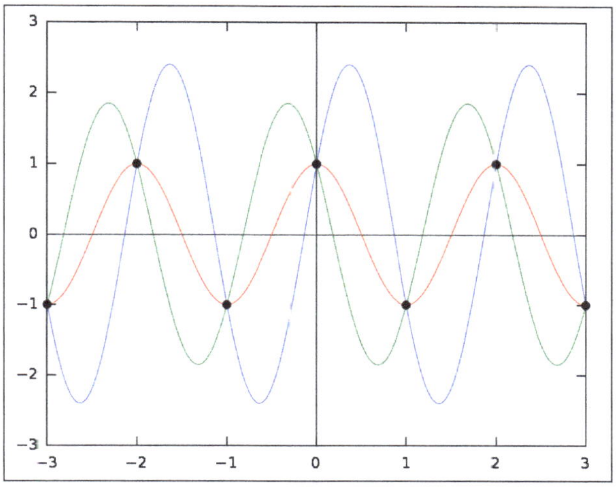

A family of sinusoids at the critical frequency, all having the same sample sequences of alternating +1 and −1. That is, they all are aliases of each other, even though their frequency is not above half the sample rate.

Sampling of Non-baseband Signals

As discussed by Shannon:

> A similar result is true if the band does not start at zero frequency but at some higher value, and can be proved by a linear translation (corresponding physically to single-sideband modulation) of the zero-frequency case. In this case the elementary pulse is obtained from $\sin(x)/x$ by single-side-band modulation.

That is, a sufficient no-loss condition for sampling signals that do not have baseband components exists that involves the *width* of the non-zero frequency interval as opposed to its highest frequency component.

For example, in order to sample the FM radio signals in the frequency range of 100–102 MHz, it is not necessary to sample at 204 MHz (twice the upper frequency), but rather it is sufficient to sample at 4 MHz (twice the width of the frequency interval).

A bandpass condition is that $X(f) = 0$, for all nonnegative f outside the open band of frequencies:

$$\left(\frac{N}{2} f_s, \frac{N+1}{2} f_s\right),$$

for some nonnegative integer N. This formulation includes the normal baseband condition as the case N=0.

The corresponding interpolation function is the impulse response of an ideal brick-wall bandpass filter (as opposed to the ideal brick-wall lowpass filter used above) with cutoffs at the upper and lower edges of the specified band, which is the difference between a pair of lowpass impulse responses:

$$(N+1)\operatorname{sinc}\left(\frac{(N+1)t}{T}\right) - N\operatorname{sinc}\left(\frac{Nt}{T}\right).$$

Other generalizations, for example to signals occupying multiple non-contiguous bands, are possible as well. Even the most generalized form of the sampling theorem does not have a provably true converse. That is, one cannot conclude that information is necessarily lost just because the conditions of the sampling theorem are not satisfied; from an engineering perspective, however, it is generally safe to assume that if the sampling theorem is not satisfied then information will most likely be lost.

Nonuniform Sampling

The sampling theory of Shannon can be generalized for the case of nonuniform sampling, that is, samples not taken equally spaced in time. The Shannon sampling theory for non-uniform sampling states that a band-limited signal can be perfectly reconstructed from its samples if the average sampling rate satisfies the Nyquist condition. Therefore, although uniformly spaced samples may result in easier reconstruction algorithms, it is not a necessary condition for perfect reconstruction.

The general theory for non-baseband and nonuniform samples was developed in 1967 by Henry Landau. He proved that the average sampling rate (uniform or otherwise) must be twice the *occupied* bandwidth of the signal, assuming it is *a priori* known what portion of the spectrum was occupied. In the late 1990s, this work was partially extended to cover signals of when the amount of occupied bandwidth was known, but the actual occupied portion of the spectrum was unknown. In the 2000s, a complete theory was developed using compressed sensing. In particular, the theory, using signal processing language, is described in this 2009 paper. They show, among other things, that if the frequency locations are unknown, then it is necessary to sample at least at twice the Nyquist criteria; in other words, you must pay at least a factor of 2 for not knowing the location of the spectrum. Note that minimum sampling requirements do not necessarily guarantee stability.

Sampling below the Nyquist Rate under Additional Restrictions

The Nyquist–Shannon sampling theorem provides a sufficient condition for the sampling and reconstruction of a band-limited signal. When reconstruction is done via the Whittaker–Shannon interpolation formula, the Nyquist criterion is also a necessary condition to avoid aliasing, in the sense that if samples are taken at a slower rate than twice the band limit, then there are some signals that will not be correctly reconstructed. However, if further restrictions are imposed on the signal, then the Nyquist criterion may no longer be a necessary condition.

A non-trivial example of exploiting extra assumptions about the signal is given by the recent field of compressed sensing, which allows for full reconstruction with a sub-Nyquist sampling rate. Specifically, this applies to signals that are sparse (or compressible) in some domain. As an example, compressed sensing deals with signals that may have a low over-all bandwidth (say, the *effective* bandwidth EB), but the frequency locations are unknown, rather than all together in a single band, so that the passband technique does not apply. In other words, the frequency spectrum is sparse. Traditionally, the necessary sampling rate is thus $2B$. Using compressed sensing techniques, the signal could be perfectly reconstructed if it is sampled at a rate slightly lower than $2EB$. With this approach, reconstruction is no longer given by a formula, but instead by the solution to a linear optimization program.

Another example where sub-Nyquist sampling is optimal arises under the additional constraint that the samples are quantized in an optimal manner, as in a combined system of sampling and optimal lossy compression. This setting is relevant in cases where the joint effect of sampling and quantization is to be considered, and can provide a lower bound for the minimal reconstruction error that can be attained in sampling and quantizing a random signal. For stationary Gaussian random signals, this lower bound is usually attained at a sub-Nyquist sampling rate, indicating that sub-Nyquist sampling is optimal for this signal model under optimal quantization.

Oversampling

In signal processing, oversampling is the process of sampling a signal at a sampling frequency significantly higher than the Nyquist rate. Theoretically, a bandwidth-limited signal can be perfectly reconstructed if sampled at the Nyquist rate or above it. The Nyquist rate is defined as twice the highest frequency component in the signal. Oversampling is capable of improving resolution and signal-to-noise ratio, and can be helpful in avoiding aliasing and phase distortion by relaxing anti-aliasing filter performance requirements.

A signal is said to be oversampled by a factor of *N* if it is sampled at *N* times the Nyquist rate.

There are three main reasons for performing oversampling:

Anti-aliasing

Oversampling can make it easier to realize analog anti-aliasing filters. Without oversampling, it is very difficult to implement filters with the sharp cutoff necessary to maximize use of the available bandwidth without exceeding the Nyquist limit. By increasing the bandwidth of the sampling system, design constraints for the anti-aliasing filter may be relaxed. Once sampled, the signal can be digitally filtered and downsampled to the desired sampling frequency. In modern integrated circuit technology, the digital filter associated with this downsampling are easier to implement than a comparable analog filter required by a non-oversampled system.

Resolution

In practice, oversampling is implemented in order to reduce cost and improve performance of an analog-to-digital converter (ADC) or digital-to-analog converter (DAC). When oversampling by a factor of N, the dynamic range also increases a factor of N because there are N times as many possible values for the sum. However, the signal-to-noise ratio (SNR) increases by $\sqrt{N}$, because summing up uncorrelated noise increases its amplitude by $\sqrt{N}$, while summing up a coherent signal increases its average by N. As a result, the SNR increases by $\sqrt{N}$.

For instance, to implement a 24-bit converter, it is sufficient to use a 20-bit converter that can run at 256 times the target sampling rate. Combining 256 consecutive 20-bit samples can increase the SNR by a factor of 16, effectively adding 4 bits to the resolution and producing a single sample with 24-bit resolution. While with N=256 there is an increase in dynamic range by 8 bits, and the level of coherent signal increases by a factor of N, the noise changes by a factor of $\sqrt{N}$ =16, so the net SNR improves by a factor of 16, 4 bits or 24 dB.

The number of samples required to get n bits of additional data precision is,

$$\text{number of samples} = (2^n)^2 = 2^{2n}.$$

To get the mean sample scaled up to an integer with 2^{2n} additional bits, the sum of samples is divided by 2^n:

$$\text{scaled mean} = \frac{\sum_{i=0}^{2^{2n}-1} 2^n \text{data}_i}{2^{2n}} = \frac{\sum_{i=0}^{2^{2n}-1} \text{data}_i}{2^n}.$$

This averaging is only effective if the signal contains sufficient uncorrelated noise to be recorded by the ADC. If not, in the case of a stationary input signal, all 2^n samples would have the same value and the resulting average would be identical to this value; so in this case, oversampling would have made no improvement. In similar cases where the ADC records no noise and the input signal is changing over time, oversampling improves the result, but to an inconsistent and unpredictable extent.

Adding some dithering noise to the input signal can actually improve the final result because the dither noise allows oversampling to work to improve resolution. In many practical applications, a small increase in noise is well worth a substantial increase in measurement resolution. In practice, the dithering noise can often be placed outside the frequency range of interest to the measurement, so that this noise can be subsequently filtered out in the digital domain—resulting in a final measurement, in the frequency range of interest, with both higher resolution and lower noise.

Noise

If multiple samples are taken of the same quantity with uncorrelated noise added to each sample, then because, uncorrelated signals combine more weakly than correlated ones, averaging *N* samples reduces the noise power by a factor of *N*. If, for example, we oversample by a factor of 4, the signal-to-noise ratio in terms of power improves by factor of 4 which corresponds to a factor of 2 improvement in terms of voltage.

Certain kinds of ADCs known as delta-sigma converters produce disproportionately more quantization noise at higher frequencies. By running these converters at some multiple of the target sampling rate, and low-pass filtering the oversampled signal down to half the target sampling rate, a final result with *less* noise (over the entire band of the converter) can be obtained. Delta-sigma converters use a technique called noise shaping to move the quantization noise to the higher frequencies.

Example:

Consider a signal with a bandwidth or highest frequency of B = 100 Hz. The sampling theorem states that sampling frequency would have to be greater than 200 Hz. Sampling at four times that rate requires a sampling frequency of 800 Hz. This gives the anti-aliasing filter a transition band of 300 Hz ($(f_s/2) - B$ = (800 Hz/2) – 100 Hz = 300 Hz) instead of 0 Hz if the sampling frequency was 200 Hz. Achieving an anti-aliasing filter with 0 Hz transition band is unrealistic whereas an anti-aliasing filter with a transition band of 300 Hz is not difficult.

Oversampling in Reconstruction

The term oversampling is also used to denote a process used in the reconstruction phase of digital-to-analog conversion, in which an intermediate high sampling rate is used between the digital input and the analogue output. Here, samples are interpolated in the digital domain to add additional samples in between, thereby converting the data to a higher sample rate, which is a form of upsampling. When the resulting higher-rate samples are converted to analog, a less complex/ expensive analog low pass filter is required to remove the high-frequency content, which will consist of reflected images of the real signal created by the zero-order hold of the digital-to-analog converter. Essentially, this is a way to shift some of the complexity of the filtering into the digital domain and achieves the same benefit as oversampling in analog-to-digital conversion.

Sampling of Continuous Signals

A continuous signal $x(t)$ can be sampled by an impulse train (or comb) function:

$$p(t) = \sum_{m=-\infty}^{\infty} \delta(t - mT_s) = \sum_{m=-\infty}^{\infty} \delta(t - \frac{m}{f_s})$$

where T_s is the sampling period, the time interval between two consecutive samples, and $f_s \overset{\Delta}{=} 1/T_s$ is the sampling rate or sampling frequency (or $\omega_s = 2\pi f_s$). The sampling process is represented by:

$$x_s(t) = x(t)\ p(t) = x(t) \sum_{m=-\infty}^{\infty} \delta(t - mT_s) = \sum_{m=-\infty}^{\infty} x(mT_s)\delta(t - mT_s)$$

where $x_s(t)$ is a train of impulses modulated by the signal $x(t)$. By the convolution theorem, the Fourier transform of $x_s(t)$ is,

$$X_s(\omega) = X(\omega) * P(\omega)$$

where $X(\omega)$ and $P(\omega)$ are the spectra of $x_s(t)$ and $p(t)$, respectively.

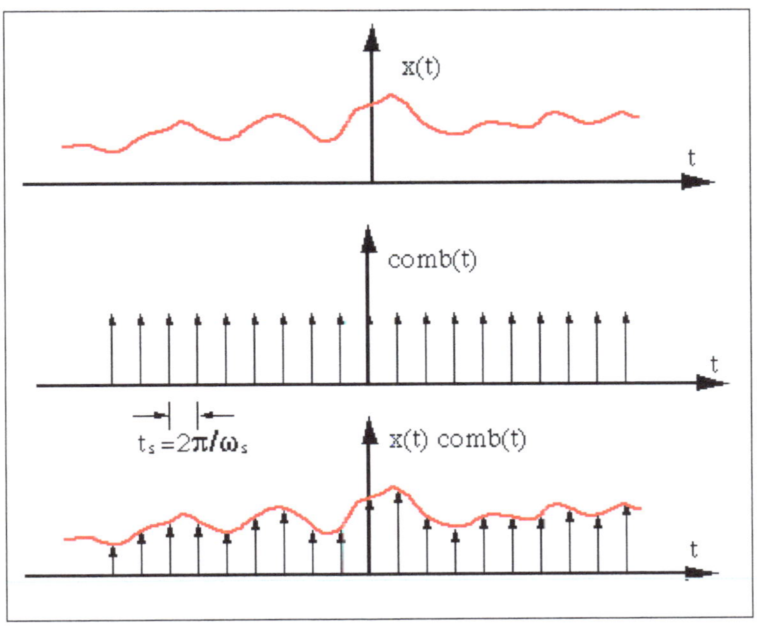

To obtain $X_s(\omega)$, we need first find out $P(\omega)$. The comb function can be treated as a special case of continuous function with period T_s and its Fourier expansion coefficients can be found to be:

$$P[k]=\frac{1}{T_s}\int_{T_s/2}^{T_s/2} x(t)e^{-j\omega_s t}dt=\frac{1}{T_s}\int_{T_s/2}^{T_s/2}\delta(t)e^{-j\omega_s t}dt=\frac{1}{T_s}$$

The spectrum of a periodic function can be obtained as a sequence of impulses weighted by its Fourier coefficients, i.e.,

$$P(\omega)=2\pi\sum_{k=-\infty}^{\infty}P[k]\,\delta(\omega-k\omega_s)=\frac{2\pi}{T_s}\sum_{k=-\infty}^{\infty}\delta(\omega-k\omega_s)$$

Also, the spectrum of the comb function $p(t)$ can be found by Fourier transform:

$$P(\omega)=\int_{-\infty}^{\infty}p(t)e^{-j\omega t}dt=\int_{-\infty}^{\infty}[\sum_{m=-\infty}^{\infty}\delta(t-mT_s)]e^{-j\omega t}dt=\sum_{m=-\infty}^{\infty}e^{-j\omega mT_s}$$

From the above two expressions of $P(\omega)$, we get the following identity as a by-product,

$$\sum_{m=-\infty}^{\infty}e^{-j\omega mT_s}=\frac{2\pi}{T_s}\sum_{k=-\infty}^{\infty}\delta(\omega-k\omega_s)$$

Given $P(\omega)$, the spectrum of the sampled signal $x_s(t)$ can be found to be,

$$X_s(\omega)\ :\ X(\omega)*P(\omega)=X(\omega)*\frac{2\pi}{T_s}\sum_{k=-\infty}^{\infty}\delta(\omega-k\omega_s)$$

$$:\ \frac{2\pi}{T_s}\sum_{k=-\infty}^{\infty}\frac{1}{2\pi}\int_{-\infty}^{\infty}X(\omega-\omega')\delta(\omega'-k\omega_s)d\omega'$$

$$:\ \frac{1}{T_s}\sum_{k=-\infty}^{\infty}X(\omega-k\omega_s)=\frac{1}{T_s}\sum_{k=-\infty}^{\infty}X(\omega-2k\pi/T_s)$$

We see that $X_s(\omega)$ is a periodic function composed of an infinite number of replicas of $X(\omega)$ repeating at period $\omega_s=2\pi/T_s$. Also note that these replicas are scaled by a factor $1/T_s$.

This figure shows the spectra of the sampled version of a band-limited signal (with ω_m as the highest frequency component in the signal) corresponding to two cases: (a) $\omega_s>2\omega_m$ and (b) $\omega_s<2\omega_m$.

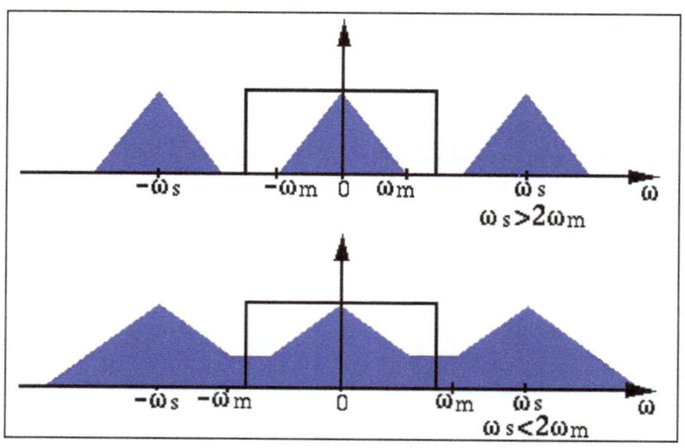

Aliasing

In signal processing and related disciplines, aliasing is an effect that causes different signals to become indistinguishable (or *aliases* of one another) when sampled. It also often refers to the distortion or artifact that results when a signal reconstructed from samples is different from the original continuous signal.

Aliasing can occur in signals sampled in time, for instance digital audio, and is referred to as temporal aliasing. It can also occur in spatially sampled signals (e.g. moiré patterns in digital images); this type of aliasing is called spatial aliasing.

Aliasing is generally avoided by applying low pass filters or anti-aliasing filters (AAF) to the input signal before sampling and when converting a signal from a higher to a lower sampling rate. Suitable reconstruction filtering should then be used when restoring the sampled signal to the continuous domain or converting a signal from a lower to a higher sampling rate. For spatial anti-aliasing, the types of anti-aliasing include fast sample anti-aliasing (FSAA), multisample anti-aliasing, and supersampling.

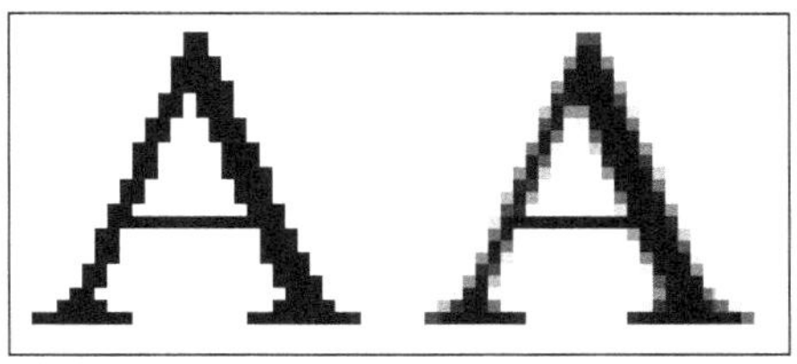

Left: An aliased image of the letter *A* in Times New Roman. Right: An *anti-aliased* image.

When a digital image is viewed, a reconstruction is performed by a display or printer device, and by the eyes and the brain. If the image data is processed in some way during sampling or reconstruction, the reconstructed image will differ from the original image, and an alias is seen.

An example of spatial aliasing is the moiré pattern observed in a poorly pixelized image of a brick wall. Spatial anti-aliasing techniques avoid such poor pixelizations. Aliasing can be caused either by the sampling stage or the reconstruction stage; these may be distinguished by calling sampling aliasing *prealiasing* and reconstruction aliasing *postaliasing*.

Temporal aliasing is a major concern in the sampling of video and audio signals. Music, for instance, may contain high-frequency components that are inaudible to humans. If a piece of music is sampled at 32000 samples per second (Hz), any frequency components at or above 16000 Hz (the Nyquist frequency for this sampling rate) will cause aliasing when the music is reproduced by a digital-to-analog converter (DAC). To prevent this, an anti-aliasing filter is used to remove components above the Nyquist frequency prior to sampling.

In video or cinematography, temporal aliasing results from the limited frame rate, and causes the wagon-wheel effect, whereby a spoked wheel appears to rotate too slowly or even backwards. Aliasing has changed its apparent frequency of rotation. A reversal of direction can be described as a negative frequency. Temporal aliasing frequencies in video and cinematography are determined by the frame rate of the camera, but the relative intensity of the aliased frequencies is determined by the shutter timing (exposure time) or the use of a temporal aliasing reduction filter during filming.

Like the video camera, most sampling schemes are periodic; that is, they have a characteristic sampling frequency in time or in space. Digital cameras provide a certain number of samples (pixels) per degree or per radian, or samples per mm in the focal plane of the camera. Audio signals are sampled (digitized) with an analog-to-digital converter, which produces a constant number of samples per second. Some of the most dramatic and subtle examples of aliasing occur when the signal being sampled also has periodic content.

Bandlimited Functions

Actual signals have a finite duration and their frequency content, as defined by the Fourier transform, has no upper bound. Some amount of aliasing always occurs when such functions are sampled. Functions whose frequency content is bounded (*bandlimited*) have an infinite duration in the time domain. If sampled at a high enough rate, determined by the *bandwidth*, the original function can, in theory, be perfectly reconstructed from the infinite set of samples.

Bandpass Signals

Sometimes aliasing is used intentionally on signals with no low-frequency content, called *bandpass* signals. Undersampling, which creates low-frequency aliases, can produce the same result, with less effort, as frequency-shifting the signal to lower frequencies before sampling at the lower rate. Some digital channelizers exploit aliasing in this way for computational efficiency.

Sampling Sinusoidal Functions

Sinusoids are an important type of periodic function, because realistic signals are often modeled as the summation of many sinusoids of different frequencies and different amplitudes (for example, with a Fourier series or transform). Understanding what aliasing does to the individual sinusoids is useful in understanding what happens to their sum.

When sampling a function at frequency f_s (intervals $1/f_s$), the following functions yield identical sets of samples: $\{\sin(2\pi(f+Nf_s)\,t + \varphi), N = 0, \pm1, \pm2, \pm3, \ldots\}$. A frequency spectrum of the samples produces equally strong responses at all those frequencies. Without collateral information, the frequency of the original function is ambiguous. So the functions and their frequencies are said to be *aliases* of each other. Noting the trigonometric identity:

$$\sin(2\pi(f+Nf_s)t+\phi) = \begin{cases} +\sin(2\pi(f+Nf_s)t+\phi), & f+Nf_s \geq 0 \\ -\sin(2\pi\,|f+Nf_s|\,t-\phi), & f+Nf_s < 0 \end{cases}$$

we can write all the alias frequencies as positive values: $\boldsymbol{f_N(f) \triangleq |f+Nf_s|}$.

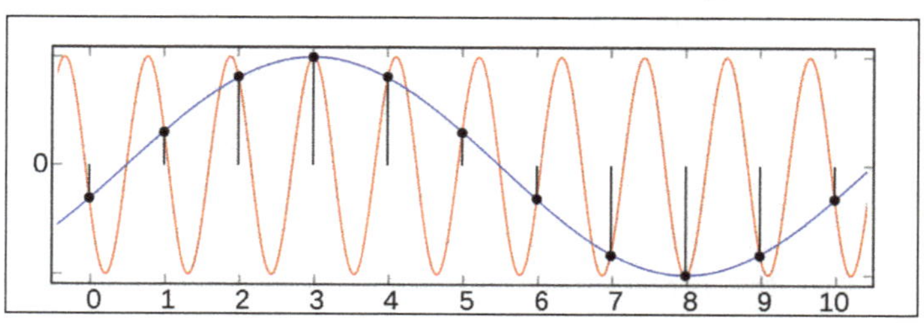

Two different sinusoids that fit the same set of samples.

For example, here a plot depicts a set of samples with parameter $f_s = 1$, and two different sinusoids that could have produced the samples. Nine cycles of the red sinusoid and one cycle of the blue sinusoid span an interval of 10 samples. The corresponding number of *cycles per sample* are $f_{red} = 0.9f_s$ and $f_{blue} = 0.1f_s$. So the $N = -1$ alias of f_{red} is f_{blue} (and vice versa).

Aliasing matters when one attempts to reconstruct the original waveform from its samples. The most common reconstruction technique produces the smallest of the $f_N(f)$ frequencies. So it is usually important that $f_0(f)$ be the unique minimum. A necessary and sufficient condition for that is $f_s/2 > |f|$, where $f_s/2$ is commonly called the Nyquist frequency of a system that samples at rate f_s. In our example, the Nyquist condition is satisfied if the original signal is the blue sinusoid ($f = f_{blue}$). But if $f = f_{red} = 0.9f_s$, the usual reconstruction method will produce the blue sinusoid instead of the red one.

Folding

In the example above, f_{red} and f_{blue} are symmetrical around the frequency $f_s/2$. And in general, as f increases from 0 to $f_s/2$, $f_{-1}(f)$ decreases from f_s to $f_s/2$. Similarly, as f increases from $f_s/2$ to f_s, $f_{-1}(f)$ continues decreasing from $f_s/2$ to 0.

A graph of amplitude vs frequency for a single sinusoid at frequency $0.6f_s$ and some of its aliases at $0.4f_s$, $1.4f_s$, and $1.6f_s$ would look like the 4 black dots in the first figure below. The red lines depict the paths (loci) of the 4 dots if we were to adjust the frequency and amplitude of the sinusoid along the solid red segment (between $f_s/2$ and f_s). No matter what function we choose to change the amplitude vs frequency, the graph will exhibit symmetry between 0 and f_s. This symmetry is commonly referred to as folding, and another name for $f_s/2$ (the Nyquist frequency) is folding frequency. Folding is often observed in practice when viewing the frequency spectrum of real-valued samples, such as the second figure below.

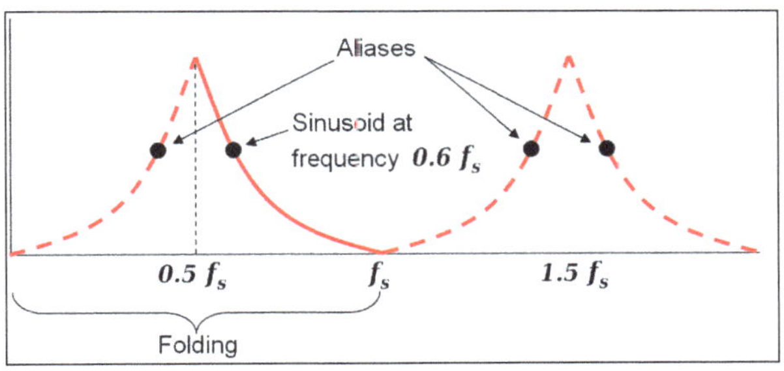

The black dots are aliases of each other. The solid red line is an example of amplitude varying with frequency. The dashed red lines are the corresponding paths of the aliases.

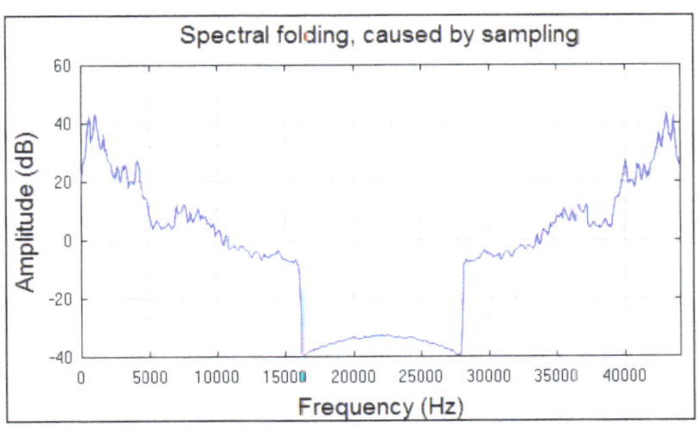

The Fourier transform of music sampled at 44100 samples/sec exhibits symmetry (called "folding") around the Nyquist frequency (22050 Hz).

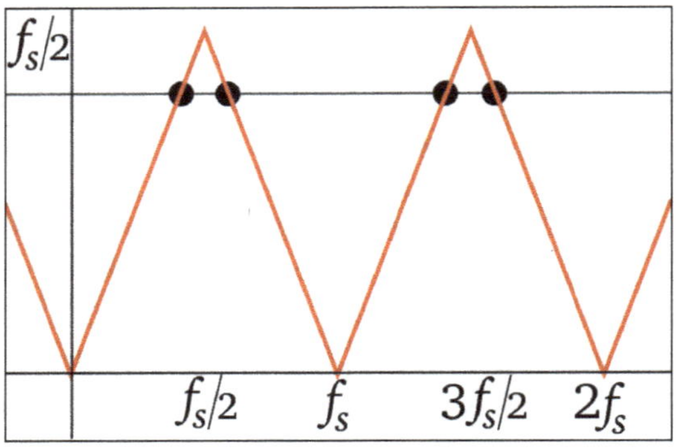

Graph of frequency aliasing, showing folding frequency and periodicity. Frequencies above $f_s/2$ have an *alias* below $f_s/2$, whose value is given by this graph.

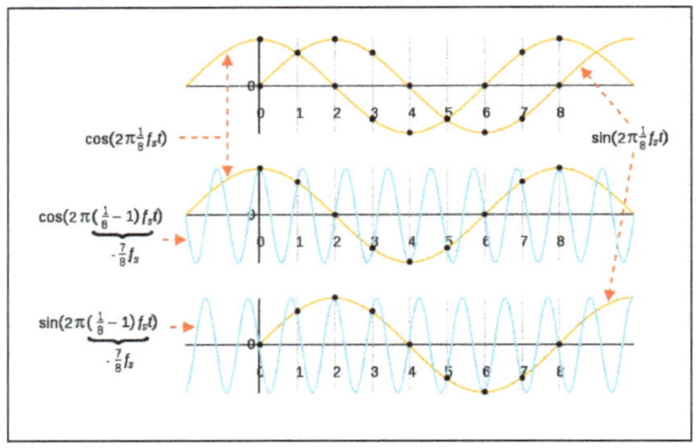

Two complex sinusoids, colored gold and cyan, that fit the same sets of real and imaginary sample points when sampled at the rate (f_s) indicated by the grid lines. The case shown here is: $f_{\text{cyan}} = f_{-1}(f_{\text{gold}}) = f_{\text{gold}} - f_s$.

Complex Sinusoids

Complex sinusoids are waveforms whose samples are complex numbers, and the concept of negative frequency is necessary to distinguish them. In that case, the frequencies of the aliases are given by just: $f_N(f) = f + Nf_s$. Therefore, as f increases from $f_s/2$ to f_s, $f_{-1}(f)$ goes from $-f_s/2$ up to 0. Consequently, complex sinusoids do not exhibit *folding*. Complex samples of real-valued sinusoids have zero-valued imaginary parts and do exhibit folding.

Sample Frequency

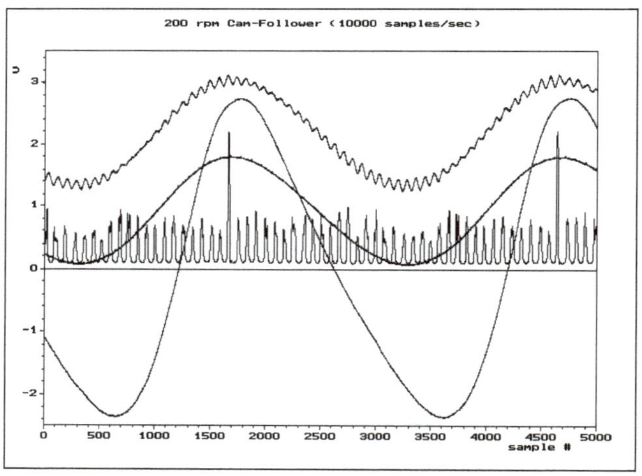

Illustration of 4 waveforms reconstructed from samples taken at six different rates. Two of the waveforms are sufficiently sampled to avoid aliasing at all six rates. The other two illustrate increasing distortion (aliasing) at the lower rates.

When the condition $f_s/2 > f$ is met for the highest frequency component of the original signal, then it is met for all the frequency components, a condition called the Nyquist criterion. That is typically approximated by filtering the original signal to attenuate high frequency components before it is sampled. These attenuated high frequency components still generate low-frequency aliases, but typically at low enough amplitudes that they do not cause problems. A filter chosen in anticipation of a certain sample frequency is called an anti-aliasing filter.

The filtered signal can subsequently be reconstructed, by interpolation algorithms, without significant additional distortion. Most sampled signals are not simply stored and reconstructed. But the fidelity of a theoretical reconstruction (via the Whittaker–Shannon interpolation formula) is a customary measure of the effectiveness of sampling.

Usage

Historically the term *aliasing* evolved from radio engineering because of the action of superheterodyne receivers. When the receiver shifts multiple signals down to lower frequencies, from RF to IF by heterodyning, an unwanted signal, from an RF frequency equally far from the local oscillator (LO) frequency as the desired signal, but on the wrong side of the LO, can end up at the same IF frequency as the wanted one. If it is strong enough it can interfere with reception of the desired signal. This unwanted signal is known as an *image* or *alias* of the desired signal.

Angular Aliasing

Aliasing occurs whenever the use of discrete elements to capture or produce a continuous signal causes frequency ambiguity.

Spatial aliasing, particular of angular frequency, can occur when reproducing a light field or sound field with discrete elements, as in 3D displays or wave field synthesis of sound.

This aliasing is visible in images such as posters with lenticular printing: if they have low angular resolution, then as one moves past them, say from left-to-right, the 2D image does not initially change (so it appears to move left), then as one moves to the next angular image, the image suddenly changes – and the frequency and amplitude of this side-to-side movement corresponds to the angular resolution of the image (and, for frequency, the speed of the viewer's lateral movement), which is the angular aliasing of the 4D light field.

The lack of parallax on viewer movement in 2D images and in 3-D film produced by stereoscopic glasses (in 3D films the effect is called "yawing", as the image appears to rotate on its axis) can similarly be seen as loss of angular resolution, all angular frequencies being aliased to 0 (constant).

Online Audio Example

The qualitative effects of aliasing can be heard in the following audio demonstration. Six sawtooth waves are played in succession, with the first two sawtooths having a fundamental frequency of 440 Hz (A4), the second two having fundamental frequency of 880 Hz (A5), and the final two at 1760 Hz (A6). The sawtooths alternate between bandlimited (non-aliased) sawtooths and aliased sawtooths and the sampling rate is 22.05 kHz. The bandlimited sawtooths are synthesized from the sawtooth waveform's Fourier series such that no harmonics above the Nyquist frequency are present.

The aliasing distortion in the lower frequencies is increasingly obvious with higher fundamental frequencies, and while the bandlimited sawtooth is still clear at 1760 Hz, the aliased sawtooth is degraded and harsh with a buzzing audible at frequencies lower than the fundamental.

Direction Finding

A form of spatial aliasing can also occur in antenna arrays or microphone arrays used to estimate the direction of arrival of a wave signal, as in geophysical exploration by seismic waves. Waves must be sampled at more than two points per wavelength, or the wave arrival direction becomes ambiguous.

Signal Reconstruction

In signal processing, reconstruction usually means the determination of an original continuous signal from a sequence of equally spaced samples.

Let *F* be any sampling method, i.e. a linear map from the Hilbert space of square-integrable functions L^2. to complex space $\mathbb{C}^n$.

In our example, the vector space of sampled signals $\mathbb{C}^n$ is *n*-dimensional complex space. Any proposed inverse *R* of *F* (*reconstruction formula*, in the lingo) would have to map $\mathbb{C}^n$ to some subset of L^2. We could choose this subset arbitrarily, but if we're going to want a reconstruction formula *R* that is also a linear map, then we have to choose an *n*-dimensional linear subspace of L^2.

This fact that the dimensions have to agree is related to the Nyquist–Shannon sampling theorem.

The elementary linear algebra approach works here. Let $d_k := (0,...,0,1,0,...,0)$ (all entries zero, except for the *k*th entry, which is a one) or some other basis of $\mathbb{C}^n$ To define an inverse for *F*, simply choose, for each *k*, an $e_k \in L^2$ so that $F(e_k) = d_k$. This uniquely defines the (pseudo-)inverse of *F*.

Of course, one can choose some reconstruction formula first, then either compute some sampling algorithm from the reconstruction formula, or analyze the behavior of a given sampling algorithm with respect to the given formula.

Ideally, the reconstruction formula is derived by minimizing the expected error variance. This requires that either the signal statistics is known or a prior probability for the signal can be specified. Information field theory is then an appropriate mathematical formalism to derive an optimal reconstruction formula.

Popular Reconstruction Formulae

Perhaps the most widely used reconstruction formula is as follows. Let $\{e_k\}$ be a basis of L^2 in the Hilbert space sense; for instance, one could use the eikonal,

$$e_k(t) := e^{2\pi ikt} \quad ,$$

although other choices are certainly possible. Note that here the index *k* can be any integer, even negative.

Then we can define a linear map R by:

$$R(d_k) = e_k$$

for each $k = \lfloor -n/2 \rfloor, \ldots, \lfloor (n-1)/2 \rfloor$, where (d_k) is the basis of $\mathbb{C}^n$ given by

$$d_k(j) = e^{\frac{2\pi ijk}{n}}$$

(This is the usual discrete Fourier basis.)

The choice of range $k = \lfloor -n/2 \rfloor, \ldots, \lfloor (n-1)/2 \rfloor$ is somewhat arbitrary, although it satisfies the dimensionality requirement and reflects the usual notion that the most important information is contained in the low frequencies. In some cases, this is incorrect, so a different reconstruction formula needs to be chosen.

A similar approach can be obtained by using wavelets instead of Hilbert bases. For many applications, the best approach is still not clear today.

Analog-to-digital Conversion

4-channel stereo multiplexed analog-to-digital converter WM8775SEDS made by Wolfson Microelectronics placed on a X-Fi Fatality Pro sound card.

In electronics, an analog-to-digital converter (ADC, A/D, or A-to-D) is a system that converts an analog signal, such as a sound picked up by a microphone or light entering a digital camera, into a digital signal. An ADC may also provide an isolated measurement such as an electronic device that converts an input analog voltage or current to a digital number representing the magnitude of the voltage or current. Typically the digital output is a two's complement binary number that is proportional to the input, but there are other possibilities.

There are several ADC architectures. Due to the complexity and the need for precisely matched components, all but the most specialized ADCs are implemented as integrated circuits (ICs).

A digital-to-analog converter (DAC) performs the reverse function; it converts a digital signal into an analog signal.

Explanation:

An ADC converts a continuous-time and continuous-amplitude analog signal to a discrete-time and discrete-amplitude digital signal. The conversion involves quantization of the input, so it necessarily introduces a small amount of error or noise. Furthermore, instead of continuously performing the conversion, an ADC does the conversion periodically, sampling the input, limiting the allowable bandwidth of the input signal.

The performance of an ADC is primarily characterized by its bandwidth and signal-to-noise ratio (SNR). The bandwidth of an ADC is characterized primarily by its sampling rate. The SNR of an ADC is influenced by many factors, including the resolution, linearity and accuracy (how well the quantization levels match the true analog signal), aliasing and jitter. The SNR of an ADC is often summarized in terms of its effective number of bits (ENOB), the number of bits of each measure it returns that are on average not noise. An ideal ADC has an ENOB equal to its resolution. ADCs are chosen to match the bandwidth and required SNR of the signal to be digitized. If an ADC operates at a sampling rate greater than twice the bandwidth of the signal, then per the Nyquist–Shannon sampling theorem, perfect reconstruction is possible. The presence of quantization error limits the SNR of even an ideal ADC. However, if the SNR of the ADC exceeds that of the input signal, its effects may be neglected resulting in an essentially perfect digital representation of the analog input signal.

Resolution

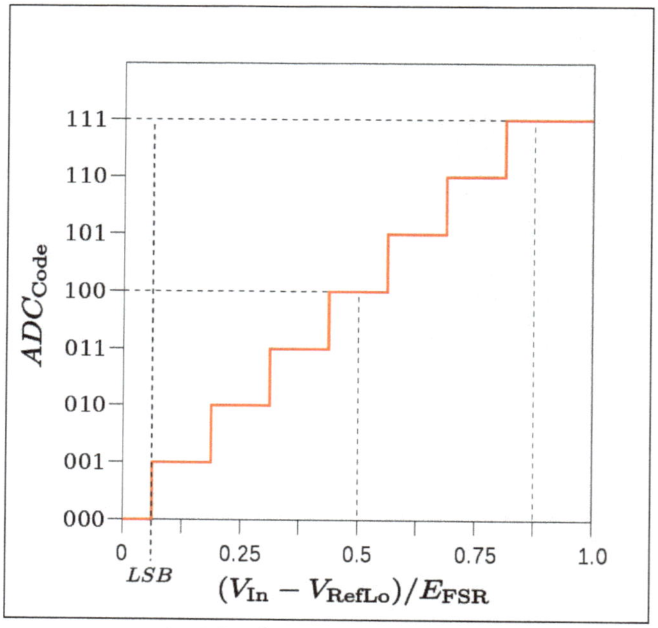

An 8-level ADC coding scheme.

The resolution of the converter indicates the number of discrete values it can produce over the range of analog values. The resolution determines the magnitude of the quantization error and therefore determines the maximum possible average signal-to-noise ratio for an ideal ADC without the use of oversampling. The values are usually stored electronically in binary form, so the resolution is usually expressed as the audio bit depth. In consequence, the number of discrete values available is assumed to be a power of two. For example, an ADC with a resolution of 8 bits can encode an analog input to one in 256 different levels ($2^8 = 256$). The values can represent the ranges from 0 to 255 (i.e. unsigned integer) or from −128 to 127 (i.e. signed integer), depending on the application.

Resolution can also be defined electrically, and expressed in volts. The change in voltage required to guarantee a change in the output code level is called the least significant bit (LSB) voltage. The resolution Q of the ADC is equal to the LSB voltage. The voltage resolution of an ADC is equal to its overall voltage measurement range divided by the number of intervals:

$$Q = \frac{E_{\mathrm{FSR}}}{2^M},$$

where M is the ADC's resolution in bits and E_{FSR} is the full scale voltage range (also called 'span'). E_{FSR} is given by:

$$E_{\mathrm{FSR}} = V_{\mathrm{RefHi}} - V_{\mathrm{RefLow}},$$

where V_{RefHi} and V_{RefLow} are the upper and lower extremes, respectively, of the voltages that can be coded.

Normally, the number of voltage intervals is given by:

$$N = 2^M,$$

where M is the ADC's resolution in bits.

That is, one voltage interval is assigned in between two consecutive code levels.

Example:

- Coding scheme as in figure
- Full scale measurement range = 0 to 1 volt
- ADC resolution is 3 bits: $2^3 = 8$ quantization levels (codes)
- ADC voltage resolution, $Q = 1\text{ V} / 8 = 0.125\text{ V}$

In many cases, the useful resolution of a converter is limited by the signal-to-noise ratio (SNR) and other errors in the overall system expressed as an ENOB.

Comparison of quantizing a sinusoid to 64 levels (6 bits) and 256 levels (8 bits). The additive noise created by 6-bit quantization is 12 dB greater than the noise created by 8-bit quantization. When the spectral distribution is flat, as in this example, the 12 dB difference manifests as a measurable difference in the noise floors.

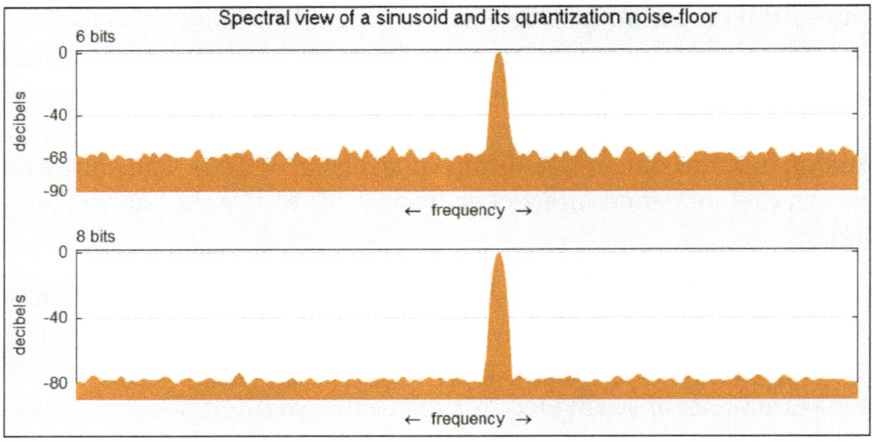

Quantization Error

Quantization error is introduced by quantization in an ideal ADC. It is a rounding error between the analog input voltage to the ADC and the output digitized value. The error is nonlinear and signal-dependent. In an ideal ADC, where the quantization error is uniformly distributed between −1/2 LSB and +1/2 LSB, and the signal has a uniform distribution covering all quantization levels, the Signal-to-quantization-noise ratio (SQNR) is given by:

$$\mathrm{SQNR} = 20\log_{10}(2^{Q}) \approx 6.02 \cdot Q\ \mathrm{dB}$$

where Q is the number of quantization bits. For example, for a 16-bit ADC, the quantization error is 96.3 dB below the maximum level.

Quantization error is distributed from DC to the Nyquist frequency, consequently if part of the ADC's bandwidth is not used, as is the case with oversampling, some of the quantization error will occur out-of-band, effectively improving the SQNR for the bandwidth in use. In an oversampled system, noise shaping can be used to further increase SQNR by forcing more quantization error out of band.

Dither

In ADCs, performance can usually be improved using dither. This is a very small amount of random noise (e.g. white noise), which is added to the input before conversion. Its effect is to randomize the state of the LSB based on the signal. Rather than the signal simply getting cut off altogether at low levels, it extends the effective range of signals that the ADC can convert, at the expense of a slight increase in noise. Note that dither can only increase the resolution of a sampler. It cannot improve the linearity, and thus accuracy does not necessarily improve.

Quantization distortion in an audio signal of very low level with respect to the bit depth of the ADC is correlated with the signal and sounds distorted and unpleasant. With dithering, the distortion is transformed into noise. The undistorted signal may be recovered accurately by averaging over time. Dithering is also used in integrating systems such as electricity meters. Since the values are added together, the dithering produces results that are more exact than the LSB of the analog-to-digital converter.

Dither is often applied when quantizing photographic images to a fewer number of bits per pixel—the image becomes noisier but to the eye looks far more realistic than the quantized image, which otherwise becomes banded. This analogous process may help to visualize the effect of dither on an analogue audio signal that is converted to digital.

Accuracy

An ADC has several sources of errors. Quantization error and (assuming the ADC is intended to be linear) non-linearity are intrinsic to any analog-to-digital conversion. These errors are measured in a unit called the least significant bit (LSB). In the above example of an eight-bit ADC, an error of one LSB is 1/256 of the full signal range, or about 0.4%.

Nonlinearity

All ADCs suffer from nonlinearity errors caused by their physical imperfections, causing their output to deviate from a linear function (or some other function, in the case of a deliberately nonlinear ADC) of their input. These errors can sometimes be mitigated by calibration, or prevented by testing. Important parameters for linearity are integral nonlinearity and differential nonlinearity. These nonlinearities introduce distortion that can reduce the signal-to-noise ratio performance of the ADC and thus reduce its effective resolution.

Jitter

When digitizing a sine wave $x(t) = A\sin(2\pi f_c t)$, the use of a non-ideal sampling clock will result in some uncertainty in when samples are recorded. Provided that the actual sampling time uncertainty due to clock jitter is Ät, the error caused by this phenomenon can be estimated as $E_{ap} \leq |x'(t)\Delta t| \leq 2A\pi f_0 \Delta t$. This will result in additional recorded noise that will reduce the effective number of bits (ENOB) below that predicted by quantization error alone. The error is zero for DC, small at low frequencies, but significant with signals of high amplitude and high frequency. The effect of jitter on performance can be compared to quantization error: Ät $< \frac{1}{2^q \pi f_0}$, where q is the number of ADC bits.

Output size (bits)	Signal Frequency						
	1 Hz	1 kHz	10 kHz	1 MHz	10 MHz	100 MHz	1 GHz
8	1,243 µs	1.24 µs	124 ns	1.24 ns	124 ps	12.4 ps	1.24 ps
10	311 µs	311 ns	31.1 ns	311 ps	31.1 ps	3.11 ps	0.31 ps
12	77.7 µs	77.7 ns	7.77 ns	77.7 ps	7.77 ps	0.78 ps	0.08 ps ("77.7fs")
14	19.4 µs	19.4 ns	1.94 ns	19.4 ps	1.94 ps	0.19 ps	0.02 ps ("19.4fs")
16	4.86 µs	4.86 ns	486 ps	4.86 ps	0.49 ps	0.05 ps ("48.5 fs")	–
18	1.21 µs	1.21 ns	121 ps	1.21 ps	0.12 ps	–	–

20	304 ns	304 ps	30.4 ps	0.30 ps ("303.56 fs")	0.03 ps ("30.3 fs")	–	–
24	18.9 ns	18.9 ps	1.89 ps	0.019 ps ("18.9 fs")	-	–	–

Clock jitter is caused by phase noise. The resolution of ADCs with a digitization bandwidth between 1 MHz and 1 GHz is limited by jitter. For lower bandwidth conversions such as when sampling audio signals at 44.1 kHz, clock jitter has a less significant impact on performance.

Sampling Rate

The analog signal is continuous in time and it is necessary to convert this to a flow of digital values. It is therefore required to define the rate at which new digital values are sampled from the analog signal. The rate of new values is called the *sampling rate* or *sampling frequency* of the converter. A continuously varying bandlimited signal can be sampled and then the original signal can be reproduced from the discrete-time values by a reconstruction filter. The Nyquist–Shannon sampling theorem implies that a faithful reproduction of the original signal is only possible if the sampling rate is higher than twice the highest frequency of the signal.

Since a practical ADC cannot make an instantaneous conversion, the input value must necessarily be held constant during the time that the converter performs a conversion (called the *conversion time*). An input circuit called a sample and hold performs this task—in most cases by using a capacitor to store the analog voltage at the input, and using an electronic switch or gate to disconnect the capacitor from the input. Many ADC integrated circuits include the sample and hold subsystem internally.

Aliasing

An ADC works by sampling the value of the input at discrete intervals in time. Provided that the input is sampled above the Nyquist rate, defined as twice the highest frequency of interest, then all frequencies in the signal can be reconstructed. If frequencies above half the Nyquist rate are sampled, they are incorrectly detected as lower frequencies, a process referred to as aliasing. Aliasing occurs because instantaneously sampling a function at two or fewer times per cycle results in missed cycles, and therefore the appearance of an incorrectly lower frequency. For example, a 2 kHz sine wave being sampled at 1.5 kHz would be reconstructed as a 500 Hz sine wave.

To avoid aliasing, the input to an ADC must be low-pass filtered to remove frequencies above half the sampling rate. This filter is called an *anti-aliasing filter*, and is essential for a practical ADC system that is applied to analog signals with higher frequency content. In applications where protection against aliasing is essential, oversampling may be used to greatly reduce or even eliminate it. Although aliasing in most systems is unwanted, it should also be noted that it can be exploited to provide simultaneous down-mixing of a band-limited high frequency signal. The alias is effectively the lower heterodyne of the signal frequency and sampling frequency.

Oversampling

Signals are often sampled at the minimum rate required, for economy, with the result that the quantization noise introduced is white noise spread over the whole pass band of the converter. If a signal is sampled at a rate much higher than the Nyquist rate and then digitally filtered to limit it to the signal bandwidth there are the following advantages:

- Digital filters can have better properties (sharper rolloff, phase) than analogue filters, so a sharper anti-aliasing filter can be realised and then the signal can be downsampled giving a better result.
- A 20-bit ADC can be made to act as a 24-bit ADC with 256× oversampling.
- The signal-to-noise ratio due to quantization noise will be higher than if the whole available band had been used. With this technique, it is possible to obtain an effective resolution larger than that provided by the converter alone.
- The improvement in SNR is 3 dB (equivalent to 0.5 bits) per octave of oversampling which is not sufficient for many applications. Therefore, oversampling is usually coupled with noise shaping . With noise shaping, the improvement is 6L+3 dB per octave where L is the order of loop filter used for noise shaping. e.g. – a 2nd order loop filter will provide an improvement of 15 dB/octave.

Oversampling is typically used in audio frequency ADCs where the required sampling rate (typically 44.1 or 48 kHz) is very low compared to the clock speed of typical transistor circuits (>1 MHz). In this case, by using the extra bandwidth to distribute quantization error onto out of band frequencies, the accuracy of the ADC can be greatly increased at no cost. Furthermore, as any aliased signals are also typically out of band, aliasing can often be completely eliminated using very low cost filters.

Relative Speed and Precision

The speed of an ADC varies by type. The Wilkinson ADC is limited by the clock rate which is processable by current digital circuits. Currently, frequencies up to 300 MHz are possible. For a successive-approximation ADC, the conversion time scales with the logarithm of the resolution, e.g. the number of bits. Thus for high resolution, it is possible that the successive-approximation ADC is faster than the Wilkinson. However, the time consuming steps in the Wilkinson are digital, while those in the successive-approximation are analog. Since analog is inherently slower than digital, as the resolution increases, the time required also increases. Thus there are competing processes at work. Flash ADCs are certainly the fastest type of the three. The conversion is basically performed in a single parallel step. For an 8-bit unit, conversion takes place in a few tens of nanoseconds.

There is, as expected, somewhat of a tradeoff between speed and precision. Flash ADCs have drifts and uncertainties associated with the comparator levels. This results in poor linearity. For successive-approximation ADCs, poor linearity is also present, but less so than for flash ADCs. Here, nonlinearity arises from accumulating errors from the subtraction processes. Wilkinson ADCs have the highest linearity of the three. These have the best differential nonlinearity. The other types require channel smoothing to achieve the level of the Wilkinson.

Sliding Scale Principle

The sliding scale or randomizing method can be employed to greatly improve the linearity of any type of ADC, but especially flash and successive approximation types. For any ADC the mapping from input voltage to digital output value is not exactly a floor or ceiling function as it should be. Under normal conditions, a pulse of a particular amplitude is always converted to a digital value. The problem lies in that the ranges of analog values for the digitized values are not all of the same width, and the differential linearity decreases proportionally with the divergence from the average width. The sliding scale principle uses an averaging effect to overcome this phenomenon. A random, but known analog voltage is added to the sampled input voltage. It is then converted to digital form, and the equivalent digital amount is subtracted, thus restoring it to its original value. The advantage is that the conversion has taken place at a random point. The statistical distribution of the final levels is decided by a weighted average over a region of the range of the ADC. This in turn desensitizes it to the width of any specific level.

Types

These are the most common ways of implementing an electronic ADC:

Direct-conversion

A direct-conversion ADC or flash ADC has a bank of comparators sampling the input signal in parallel, each firing for their decoded voltage range. The comparator bank feeds a logic circuit that generates a code for each voltage range. Direct conversion is very fast, capable of gigahertz sampling rates, but usually has only 8 bits of resolution or fewer, since the number of comparators needed, $2^N - 1$, doubles with each additional bit, requiring a large, expensive circuit. ADCs of this type have a large die size, a high input capacitance, high power dissipation, and are prone to produce glitches at the output (by outputting an out-of-sequence code). Scaling to newer submicrometre technologies does not help as the device mismatch is the dominant design limitation. They are often used for video, wideband communications or other fast signals in optical storage. There are four different types of direct ADCs.

Parallel Comparator ADC

This is the simplest ADC. It is at the same time the fastest and the most expensive technique. The circuit consists of a resistive divider network, a set of op-amp comparators and a priority encoder. A small amount of hysteresis is built into the comparator to resolve any problems that might occur if both inputs were of equal voltage. At each node of the resistive divider, a comparison voltage is available. The purpose of the circuit is to compare the analog input voltage with each of the node voltages. The circuit has the advantage of high speed as the conversion takes place simultaneously rather than sequentially. Typical conversion time is 100 ns or less. Conversion time is limited only by the speed of the comparator and of the priority encoder. This type of ADC has the disadvantage that the number of comparators required almost doubles for each added bit. Also, the larger the value of n, the more complex is the priority encoder.

Counter Type ADC

The D to A converter can be easily turned around to provide the inverse function A to D conversion. The principle is to adjust the DAC's input code until the DAC's output comes within $\pm\frac{1}{2}$ LSB to the analog input which is to be converted to binary digital form.

Servo Tracking ADC

It is an improved version of a counting ADC. The circuit consists of an up-down counter with the comparator controlling the direction of the count. The analog output of the DAC is compared with the analog input. If the input is greater than the DAC output signal, the output of the comparator goes high and the counter is caused to count up. The tracking ADC has an advantage of being simple. The disadvantage, however, is the time needed to stabilize as a new conversion value is directly proportional to the rate at which the analog signal changes.

Successive Approximation

A successive-approximation ADC uses a comparator to successively narrow a range that contains the input voltage. At each successive step, the converter compares the input voltage to the output of an internal digital to analog converter which might represent the midpoint of a selected voltage range. At each step in this process, the approximation is stored in a successive approximation register (SAR). For example, consider an input voltage of 6.3 V and the initial range is 0 to 16 V. For the first step, the input 6.3 V is compared to 8 V (the midpoint of the 0–16 V range). The comparator reports that the input voltage is less than 8 V, so the SAR is updated to narrow the range to 0–8 V. For the second step, the input voltage is compared to 4 V (midpoint of 0–8). The comparator reports the input voltage is above 4 V, so the SAR is updated to reflect the input voltage is in the range 4–8 V. For the third step, the input voltage is compared with 6 V (halfway between 4 V and 8 V); the comparator reports the input voltage is greater than 6 volts, and search range becomes 6–8 V. The steps are continued until the desired resolution is reached.

Ramp-compare

A ramp-compare ADC produces a saw-tooth signal that ramps up or down then quickly returns to zero. When the ramp starts, a timer starts counting. When the ramp voltage matches the input, a comparator fires, and the timer's value is recorded. Timed ramp converters require the fewest transistors. The ramp time is sensitive to temperature because the circuit generating the ramp is often a simple oscillator. There are two solutions: use a clocked counter driving a DAC and then use the comparator to preserve the counter's value, or calibrate the timed ramp. A special advantage of the ramp-compare system is that comparing a second signal just requires another comparator, and another register to store the voltage value. A very simple (nonlinear) ramp converter can be implemented with a microcontroller and one resistor and capacitor. Vice versa, a filled capacitor can be taken from an integrator, time-to-amplitude converter, phase detector, sample and hold circuit, or peak and hold circuit and discharged. This has the advantage that a slow comparator cannot be disturbed by fast input changes.

Wilkinson

The Wilkinson ADC was designed by D. H. Wilkinson in 1950. The Wilkinson ADC is based on the comparison of an input voltage with that produced by a charging capacitor. The capacitor is allowed to charge until its voltage is equal to the amplitude of the input pulse (a comparator determines when this condition has been reached). Then, the capacitor is allowed to discharge linearly, which produces a ramp voltage. At the point when the capacitor begins to discharge, a gate pulse is initiated. The gate pulse remains on until the capacitor is completely discharged. Thus the duration of the gate pulse is directly proportional to the amplitude of the input pulse. This gate pulse operates a linear gate which receives pulses from a high-frequency oscillator clock. While the gate is open, a discrete number of clock pulses pass through the linear gate and are counted by the address register. The time the linear gate is open is proportional to the amplitude of the input pulse, thus the number of clock pulses recorded in the address register is proportional also. Alternatively, the charging of the capacitor could be monitored, rather than the discharge.

Integrating

An integrating ADC (also dual-slope or multi-slope ADC) applies the unknown input voltage to the input of an integrator and allows the voltage to ramp for a fixed time period (the run-up period). Then a known reference voltage of opposite polarity is applied to the integrator and is allowed to ramp until the integrator output returns to zero (the run-down period). The input voltage is computed as a function of the reference voltage, the constant run-up time period, and the measured run-down time period. The run-down time measurement is usually made in units of the converter's clock, so longer integration times allow for higher resolutions. Likewise, the speed of the converter can be improved by sacrificing resolution. Converters of this type (or variations on the concept) are used in most digital voltmeters for their linearity and flexibility.

Charge Balancing ADC

The principle of charge balancing ADC is to first convert the input signal to a frequency using a voltage to frequency converter.This frequency is then measured by a counter and converted to an output code proportional to the analog input.The main advantage of these converters is that it is possible to transmit frequency even in noisy environment or in isolated form.However, the limitation of this circuit is that the output of V/F converter depends upon an RC product whose value cannot be easily maintained with temperature and time.

Dual-slope ADC

The analog part of the circuit consists of a high input impedance buffer,precision integrator and a voltage comparator.The converter first integrates the analog input signal for a fixed duration and then it integrates an internal reference voltage of opposite polarity until the integrator output is zero.The main disadvantage of this circuit is the long duration time.They are particularly suitable for accurate measurement of slowly varying signals such as thermocouples and weighing scales.

Delta-encoded

A delta-encoded ADC or counter-ramp has an up-down counter that feeds a digital to analog converter (DAC). The input signal and the DAC both go to a comparator. The comparator controls the counter. The circuit uses negative feedback from the comparator to adjust the counter until the DAC's output is close enough to the input signal. The number is read from the counter. Delta converters have very wide ranges and high resolution, but the conversion time is dependent on the input signal level, though it will always have a guaranteed worst-case. Delta converters are often very good choices to read real-world signals. Most signals from physical systems do not change abruptly. Some converters combine the delta and successive approximation approaches; this works especially well when high frequencies are known to be small in magnitude.

Pipelined

A pipelined ADC (also called subranging quantizer) uses two or more steps of subranging. First, a coarse conversion is done. In a second step, the difference to the input signal is determined with a digital to analog converter (DAC). This difference is then converted finer, and the results are combined in a last step. This can be considered a refinement of the successive-approximation ADC wherein the feedback reference signal consists of the interim conversion of a whole range of bits (for example, four bits) rather than just the next-most-significant bit. By combining the merits of the successive approximation and flash ADCs this type is fast, has a high resolution, and only requires a small die size.

Sigma-delta

A sigma-delta ADC (also known as a delta-sigma ADC) oversamples the desired signal by a large factor and filters the desired signal band. Generally, a smaller number of bits than required are converted using a Flash ADC after the filter. The resulting signal, along with the error generated by the discrete levels of the Flash, is fed back and subtracted from the input to the filter. This negative feedback has the effect of noise shaping the error due to the Flash so that it does not appear in the desired signal frequencies. A digital filter (decimation filter) follows the ADC which reduces the sampling rate, filters off unwanted noise signal and increases the resolution of the output (sigma-delta modulation, also called delta-sigma modulation).

Time-interleaved

A time-interleaved ADC uses M parallel ADCs where each ADC samples data every M:th cycle of the effective sample clock. The result is that the sample rate is increased M times compared to what each individual ADC can manage. In practice, the individual differences between the M ADCs degrade the overall performance reducing the spurious-free dynamic range (SFDR). However, technologies exist to correct for these time-interleaving mismatch errors.

Intermediate FM Stage

An ADC with intermediate FM stage first uses a voltage-to-frequency converter to convert the desired signal into an oscillating signal with a frequency proportional to the voltage of the desired signal, and then uses a frequency counter to convert that frequency into a digital count proportional to the desired signal voltage. Longer integration times allow for higher resolutions. Likewise, the speed of the converter can be improved by sacrificing resolution. The two parts of the ADC may be widely separated, with the frequency signal passed through an opto-isolator or transmitted wirelessly. Some such ADCs use sine wave or square wave frequency modulation; others use pulse-frequency modulation. Such ADCs were once the most popular way to show a digital display of the status of a remote analog sensor.

Other Types

There can be other ADCs that use a combination of electronics and other technologies. A time-stretch analog-to-digital converter (TS-ADC) digitizes a very wide bandwidth analog signal, that cannot be digitized by a conventional electronic ADC, by time-stretching the signal prior to digitization. It commonly uses a photonic preprocessor frontend to time-stretch the signal, which effectively slows the signal down in time and compresses its bandwidth. As a result, an electronic backend ADC, that would have been too slow to capture the original signal, can now capture this slowed down signal. For continuous capture of the signal, the frontend also divides the signal into multiple segments in addition to time-stretching. Each segment is individually digitized by a separate electronic ADC. Finally, a digital signal processor rearranges the samples and removes any distortions added by the frontend to yield the binary data that is the digital representation of the original analog signal.

Commercial

Commercial ADCs are usually implemented as integrated circuits. Most converters sample with 6 to 24 bits of resolution, and produce fewer than 1 megasample per second. Thermal noise generated by passive components such as resistors masks the measurement when higher resolution is desired. For audio applications and in room temperatures, such noise is usually a little less than 1 μV (microvolt) of white noise. If the MSB corresponds to a standard 2 V of output signal, this translates to a noise-limited performance that is less than 20~21 bits, and obviates the need for any dithering. Mega- and giga-sample per second converters are available. Mega-sample converters are required in digital video cameras, video capture cards, and TV tuner cards to convert full-speed analog video to digital video files. Commercial converters usually have ±0.5 to ±1.5 LSB error in their output.

In many cases, the most expensive part of an integrated circuit is the pins, because they make the package larger, and each pin has to be connected to the integrated circuit's silicon. To save pins, it is common for slow ADCs to send their data one bit at a time over a serial interface to the computer, with the next bit coming out when a clock signal changes state, say from 0 to 5 V. This saves quite a few pins on the ADC package, and in many cases, does not make the overall design any more complex (even microprocessors which use memory-mapped I/O only need a few bits of a port to implement a serial bus to an ADC). Commercial ADCs often have several inputs that feed the same converter, usually through an analog multiplexer. Different models of ADC may include sample and hold circuits, instrumentation amplifiers or differential inputs, where the quantity measured is the difference between two voltages.

Applications

Music Recording

Analog-to-digital converters are integral to 2000s era music reproduction technology and digital audio workstation-based sound recording. People often produce music on computers using an analog recording and therefore need analog-to-digital converters to create the pulse-code modulation (PCM) data streams that go onto compact discs and digital music files. The current crop of analog-to-digital converters utilized in music can sample at rates up to 192 kilohertz. Considerable literature exists on these matters, but commercial considerations often play a significant role. Many recording studios record in 24-bit/96 kHz (or higher) pulse-code modulation (PCM) or Direct Stream Digital (DSD) formats, and then downsample or decimate the signal for Compact Disc Digital Audio production (44.1 kHz) or to 48 kHz for commonly used radio and television broadcast applications because of the Nyquist frequency and hearing range of humans.

Digital Signal Processing

ADCs are required to process, store, or transport virtually any analog signal in digital form. TV tuner cards, for example, use fast video analog-to-digital converters. Slow on-chip 8, 10, 12, or 16 bit analog-to-digital converters are common in microcontrollers. Digital storage oscilloscopes need very fast analog-to-digital converters, also crucial for software defined radio and their new applications.

Scientific Instruments

Digital imaging systems commonly use analog-to-digital converters in digitizing pixels. Some radar systems commonly use analog-to-digital converters to convert signal strength to digital values for subsequent signal processing. Many other in situ and remote sensing systems commonly use analogous technology. The number of binary bits in the resulting digitized numeric values reflects the resolution, the number of unique discrete levels of quantization (signal processing). The correspondence between the analog signal and the digital signal depends on the quantization error. The quantization process must occur at an adequate speed, a constraint that may limit the resolution of the digital signal. Many sensors in scientific instruments produce an analog signal; temperature, pressure, pH, light intensity etc. All these signals can be amplified and fed to an ADC to produce a digital number proportional to the input signal.

Rotary Encoder

Some non-electronic or only partially electronic devices, such as rotary encoders, can also be considered ADCs. Typically the digital output of an ADC will be a two's complement binary number that is proportional to the input. An encoder might output a Gray code.

Electrical Symbol

Testing

Testing an Analog to Digital Converter requires an analog input source and hardware to send control signals and capture digital data output. Some ADCs also require an accurate source of reference signal.

The key parameters to test a SAR ADC are:

- DC offset error,
- DC gain error,
- Signal to noise ratio (SNR),

- Total harmonic distortion (THD),
- Integral non linearity (INL),
- Differential non linearity (DNL),
- Spurious free dynamic range,
- Power dissipation.

Digital to Analog Conversion

Digital Signal – A digital signal is a signal that represents data as a sequence of discrete values; at any given time it can only take on one of a finite number of values.

Analog Signal – An analog signal is any continuous signal for which the time varying feature of the signal is a representation of some other time varying quantity i.e., analogous to another time varying signal.

The following techniques can be used for Digital to Analog Conversion:

1. Amplitude Shift keying – Amplitude Shift Keying is a technique in which carrier signal is analog and data to be modulated is digital. The amplitude of analog carrier signal is modified to reflect binary data.

The binary signal when modulated gives a zero value when the binary data represents 0 while gives the carrier output when data is 1. The frequency and phase of the carrier signal remain constant.

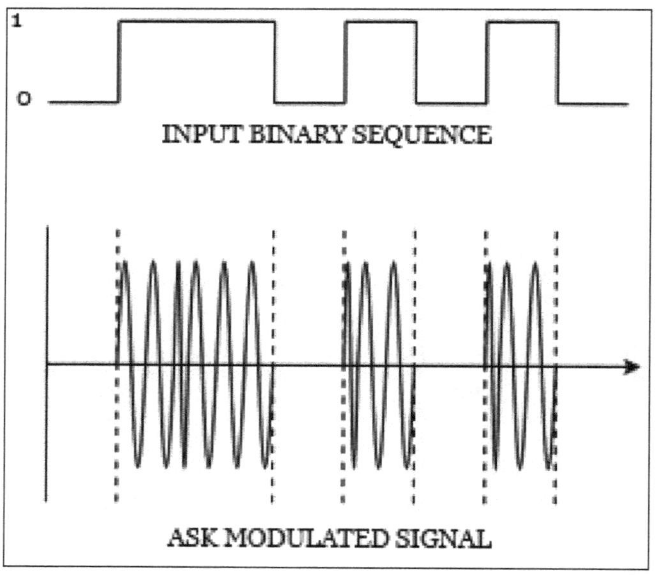

Advantages of amplitude shift Keying:

- It can be used to transmit digital data over optical fiber.

- The receiver and transmitter have a simple design which also makes it comparatively inexpensive.
- It uses lesser bandwidth as compared to FSK thus it offers high bandwidth efficiency.

Disadvantages of amplitude shift Keying:

- It is susceptible to noise interference and entire transmissions could be lost due to this.
- It has lower power efficiency.

2. Frequency Shift keying – In this modulation the frequency of analog carrier signal is modified to reflect binary data.

The output of a frequency shift keying modulated wave is high in frequency for a binary high input and is low in frequency for a binary low input. The amplitude and phase of the carrier signal remain constant.

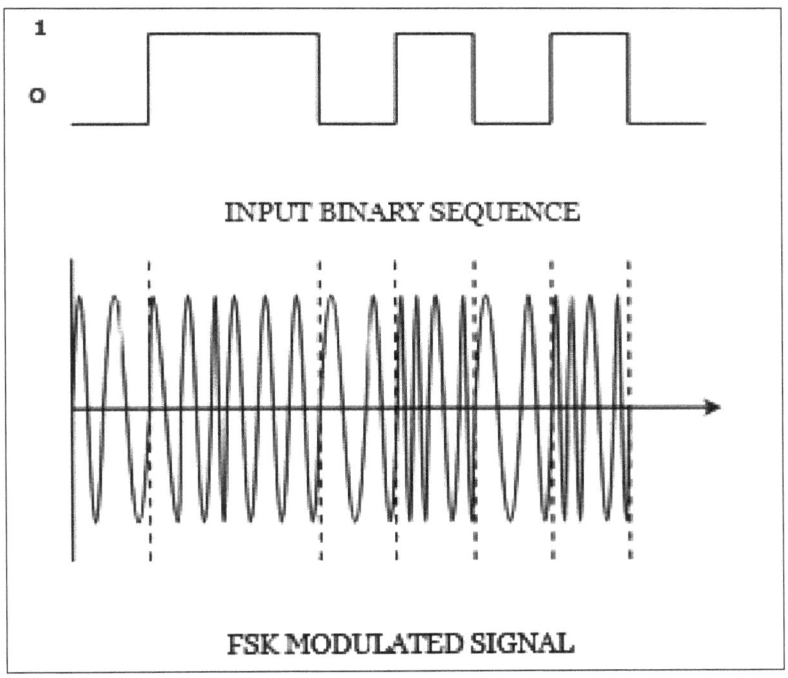

Advantages of frequency shift Keying:

- Frequency shift keying modulated signal can help avoid the noise problems beset by ASK.
- It has lower chances of an error.
- It provides high signal to noise ratio.
- The transmitter and receiver implementations are simple for low data rate application.

Disadvantages of frequency shift Keying:

- It uses larger bandwidth as compared to ASK thus it offers less bandwidth efficiency.

- It has lower power efficiency.

3. Phase Shift keying – In this modulation the phase of the analog carrier signal is modified to reflect binary data.The amplitude and frequency of the carrier signal remains constant.

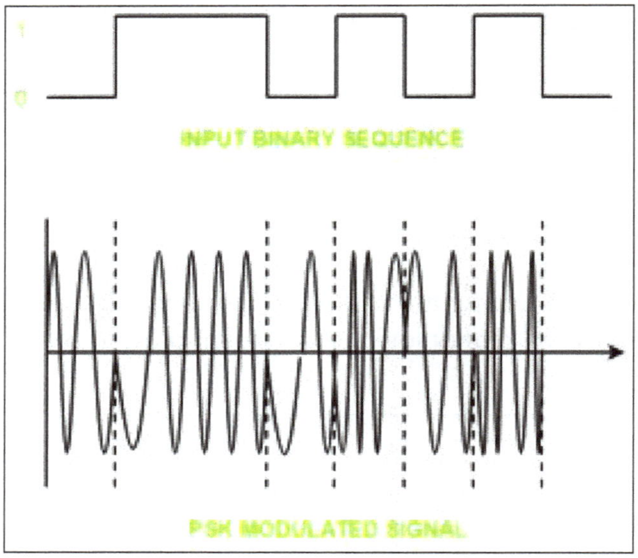

It is further categorized as follows:

- Binary Phase Shift Keying (BPSK): BPSK also known as phase reversal keying or 2PSK is the simplest form of phase shift keying. The Phase of the carrier wave is changed according to the two binary inputs. In Binary Phase shift keying, difference of 180 phase shift is used between binary 1 and binary 0.

This is regarded as the most robust digital modulation technique and is used for long distance wireless communication.

- Quadrature phase shift keying: This technique is used to increase the bit rate i.e we can code two bits onto one single element. It uses four phases to encode two bits per symbol. QPSK uses phase shifts of multiples of 90 degrees.

It has double data rate carrying capacity compare to BPSK as two bits are mapped on each constellation points.

Advantages of phase shift Keying:

- It is a more power efficient modulation technique as compared to ASK and FSK.
- It has lower chances of an error.
- It allows data to be carried along a communication signal much more efficiently as compared to FSK.

Disadvantages of phase shift Keying:

- It offers low bandwidth efficiency.

- The detection and recovery algorithms of binary data is very complex.
- It is a non coherent reference signal.

References

- Kipnis, Alon; Goldsmith, Andrea J.; Eldar, Yonina C.; Weissman, Tsachy (January 2016). "Distortion rate function of sub-Nyquist sampled Gaussian sources". IEEE Transactions on Information Theory. 62: 401–429. Arxiv:1405.5329. Doi:10.1109/tit.2015.2485271
- Holma, Tomlinson (2012). Sound for Film and Television. CRC Press. Pp. 52–53. ISBN 9781136046100. Retrieved 4 February 2019
- Digital-electronics-digital-analog-conversion: geeksforgeeks.org, Retrieved 14 July, 2019
- Nauman Uppal (30 August 2004). "Upsampling vs. Oversampling for Digital Audio". Retrieved 6 October 2012
- Harris, frederic j. (Aug 2006). Multirate Signal Processing for Communication Systems. Upper Saddle River, NJ: Prentice Hall PTR. ISBN 978-0-13-146511-4

4

Discrete Fourier Transform

Discrete Fourier transform refers to a method which is used to convert a finite sequence of functions which are equally-spaced into a same-length sequence of equally-spaced samples of the discrete-time Fourier transform. This chapter discusses discrete Fourier transform and signal spectrum in detail.

The Discrete Fourier Transform (DFT) is the equivalent of the continuous Fourier Transform for signals known only at N instants separated by sample times T (i.e. a finite sequence of data).

Let $f(t)$ be the continuous signal which is the source of the data. Let N samples be denoted,

$$f[0], f[1], f[2], \ldots, f[k], \ldots, f[N-1].$$

The Fourier Transform of the original signal, $f(t)$, would be,

$$F(j\omega) = \int_{-\infty}^{\infty} f(te^{-j\omega t} dt$$

We could regard each sample $f[k]$ as an impulse having area $f[k]$. Then, since the integrand exists only at the sample points:

$$F(j\omega) = \int_{0}^{(N-1)T} f(t)\, e^{-j\omega t} dt$$

$$= f[0]e^{-j0} + f[1]e^{-j\omega t} + \ldots + f[k]e^{-j\omega kT} + \ldots F(N-1)e^{-j\omega(N-1)T}$$

$$F(j\omega) = \sum_{k=0}^{N-1} f[k]\, e^{-j\omega kT}$$

We could in principle evaluate this for any ω, but with only N data points to start with, only N final outputs will be significant.

You may remember that the continuous Fourier transform could be evaluated over a finite interval (usually the fundamental period T_O) rather than from $-\infty$ to $+\infty$ if the waveform was periodic. Similarly, since there are only a finite number of input data points, the DFT treats the data as if it were periodic (i.e. $f(N)$ to $f(2N-1)$ is the same $f(0)$ to $(N-1)$.

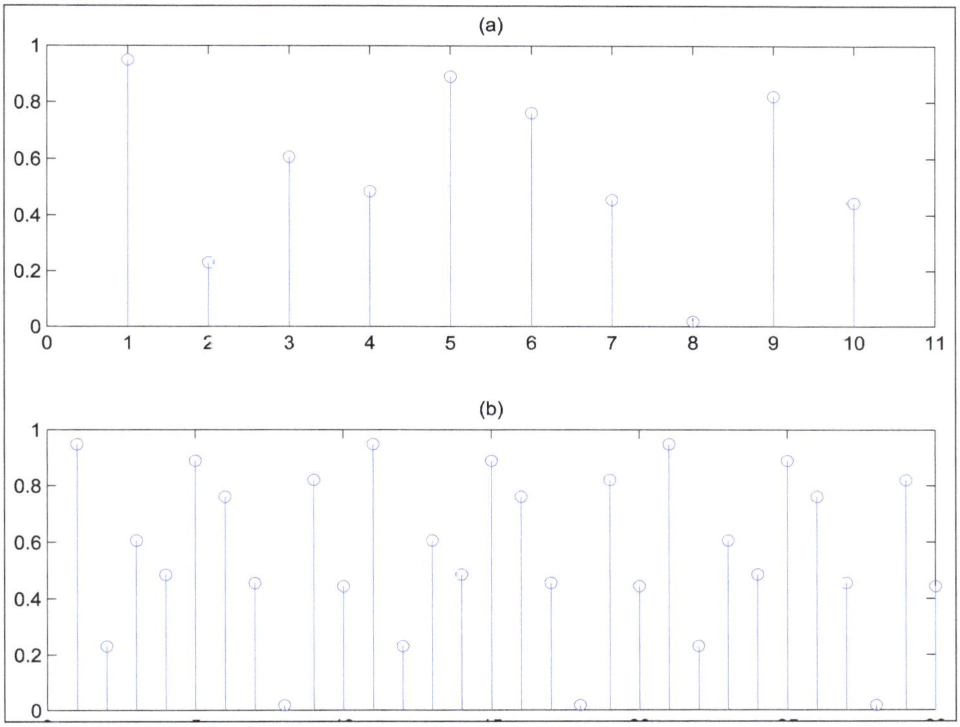

(a) Sequence of $N = 10$ samples. (b) implicit periodicity in DFT.

Since the operation treats the data as if it were periodic, we evaluate the DFT equation for the fundamental frequency (one cycle per sequence, $\frac{1}{NT}$ HZ, $\frac{2\pi}{NT}$ $rad/\text{sec.}$) and its harmonics (not forgetting the d.c. component (or average) at $\omega = 0$).

$$\text{i.e. set}\,\omega = 0, \frac{2\pi}{NT}, \frac{2\pi}{NT}\times 2, \ldots \frac{2\pi}{NT}\times n, \ldots \frac{2\pi}{NT}\times (N-1)$$

or, in general

$$F[n] = \sum_{K=0}^{N-1} f[k] e^{-j\frac{2\pi}{N}nk} \qquad (n = 0 : N-1)$$

$F[n]$ is the Discrete Fourier Transform of the sequence $f[k]$.

We may write this equation in matrix form as:

$$\begin{pmatrix} F[0] \\ F[1] \\ F[2] \\ \vdots \\ F[N-1] \end{pmatrix} = \begin{pmatrix} 1 & 1 & 1 & 1 & \ldots & 1 \\ 1 & W & W^2 & W^3 & \ldots & W^{N-1} \\ 1 & W^2 & W^4 & W^6 & \ldots & W^{N-2} \\ 1 & W^3 & W^6 & W^9 & \ldots & W^{N-3} \\ \vdots & & & & & \\ 1 & W^{N-1} & W^{N-2} & W^{N-3} & \ldots & W \end{pmatrix} \begin{pmatrix} f[0] \\ f[1] \\ f[2] \\ \vdots \\ f[N-1] \end{pmatrix}$$

Where $W = \exp(-j2\pi / N)$ and $W = W^{2N}$ etc. $= 1$.

DFT – Example

Let the continuous signal,

$$f(t) = \underbrace{5}_{dc} + \underbrace{2\cos(2\pi t - 90^o)}_{1\text{Hz}} + \underbrace{3\cos 4\pi t}_{2\text{Hz}}$$

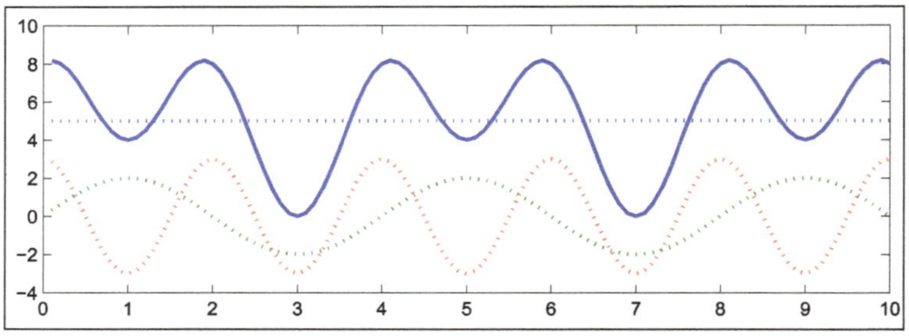

Example signal for DFT.

Let us sample $f(t)$ at 4 times per second $(ie.\ f_s = 4Hz)$ from $t = 0$ to $t = \frac{3}{4}$. The e values of the discrete samples are given by:

$$f[k] = 5 + 2\cos\left(\frac{\pi}{2}k - 90^{\text{o}}\right) + 3\cos\pi k \text{ by putting } t = kT_s = \frac{k}{4}$$

$$\text{i.e. } f[0] = 8,\ f[1] = 4,\ f[2] = 8,\ f[3] = 0, \qquad (N = 4)$$

Therefore $F[n] = \sum_0^3 f[k] e^{-j\frac{\pi}{2}nk} = \sum_{k=0}^{3} f[k](-j)^{nk}$

$$\begin{pmatrix} F[0] \\ F[1] \\ F[2] \\ F[3] \end{pmatrix} = \begin{pmatrix} 1 & 1 & 1 & 1 \\ 1 & -j & -1 & j \\ 1 & -1 & 1 & -1 \\ 1 & j & -1 & -j \end{pmatrix} \begin{pmatrix} f[0] \\ f[1] \\ f[2] \\ f[3] \end{pmatrix} = \begin{pmatrix} 20 \\ -j4 \\ 12 \\ j4 \end{pmatrix}$$

The magnitude of the DFT coefficients is shown below in figure:

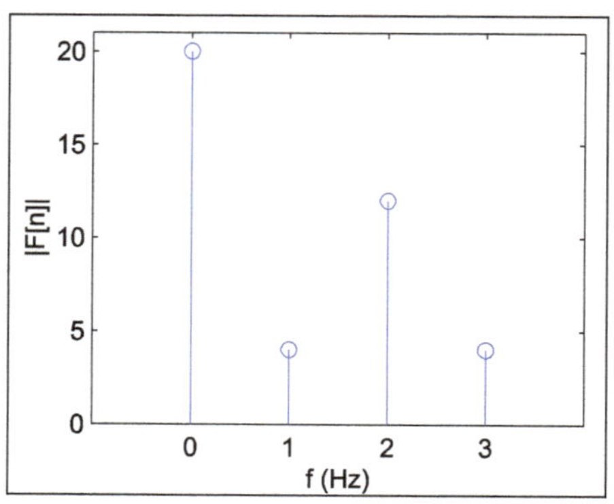

DFT of four point sequence.

Inverse Discrete Fourier Transform

The inverse transform of,

$$F[n] = \sum_{K=0}^{N-1} f[k] e^{-j\frac{2\pi}{N}nk}$$

$$f[k] = \frac{1}{N}\sum_{n=0}^{N-1} F[n] e^{+j\frac{2\pi}{N}nk}$$

i.e. the inverse matrix is $\frac{1}{N}$ times the complex conjugate of the original (symmetric) matrix.

Note that the $F[n]$ coefficients are complex. We can assume that the $f[k]$ values are real (this is the simplest case; there are situations (e.g. radar) in which two inputs, at each k, are treated as a complex pair, since they are the outputs from 0° and 90° demodulators).

In the process of taking the inverse transform the term $F[n]$ and $F[N-n]$ (remember that the e spectrum is symmetrical about $\frac{N}{2}$) combine to produce 2 frequency components, only one of which is considered to be valid (the one at the lower of the two frequencies, $n \times \frac{2\pi}{T} Hz$ where $n \le \frac{N}{2}$; the higher frequency component is at an "aliasing frequency" $(n > \frac{N}{2})$).

From the inverse transform formula, the contribution $f[k]$ of $F[n]$ and $F[N-n]$ is:

$$f_n[k] = \frac{1}{N}\{F[n] e^{j\frac{2\pi}{N}(n)k} + F[N-n] e^{j\frac{2\pi}{N}(N-n)k}\}$$

For all $f[k]$ real, $F[N-n] = \sum_{K=0}^{N-1} f[k] e^{-j\frac{2\pi}{N}(N-n)k}$

But $e^{-j\frac{2\pi}{N}(N-n)k} = \underbrace{e^{-j2\pi k}}_{1 \text{ for all } k} e^{+j\frac{2\pi n}{N}k} = e + ^{j\frac{2\pi n}{N}nk}$

i.e. $F[N-n] = F^*(n)$ (i.e. the complex conjugate)

Substituting into the Equation for $f_n[k]$ above gives,

$$f_n[k] = \frac{1}{N}\{F[n] e^{j\frac{2\pi}{N}nk} + F^*(n) e^{-j\frac{2\pi}{N}nk}\} \qquad \text{since } e^{j2\pi k} = 1$$

ie. $f_n[k] = \frac{2}{N}\{\mathrm{Re}\{F[n]\}\cos\frac{2\pi}{N}nk - \mathrm{Im}\{F[n]\}\sin\frac{2\pi}{N}nk\}$

or $f_n[k] = \frac{2}{N}|F[n]|\cos\{(\frac{2\pi}{NT}n)kT + \arg(F[n])\}$

d i.e. a sampled sinewave at $\frac{2\pi n}{NT}$ Hz, of magnitude $\frac{2}{N}|F[n]|$.

For the special case of $n=0$, $F[0]=f[k]$ i.e. sum of all samples) and the contribution of $F[0]$ to $f[k]$ is $f_0[k]=\frac{1}{N}F[0]=$ average of $f[k]=$ d.c. component.

Interpretation of Example

1. $F[0]=20$ implies a d. c. value of $\frac{1}{N}F[0]=\frac{20}{4}=5$ (as expected).
2. $F[1]=-j4=F^*[3]$ impliesafundamentalcomponentofpeakamplitude $\frac{2}{N}|F[1]|=\frac{2}{4}\times 4=2$

   with phase given by $\arg F[1]=-90^{\circ}$

   i.e. $2\cos(\frac{2\pi}{NT}kT-90^{\circ})=2\cos(\frac{\pi}{2}k-90^{\circ})$ (as expected).

3. $F[2]=12$ $(n=\frac{N}{2}$ – no other $N-n$ component here$)$ and this implies a component

   $$f_2[k]=\frac{1}{N}F[2]e^{j\frac{2\pi}{N}\cdot 2K}=\frac{1}{4}F[2]e^{j\pi K}=3\cos\pi k$$ '(as expected)

   sincesinπ k = 0for all k.

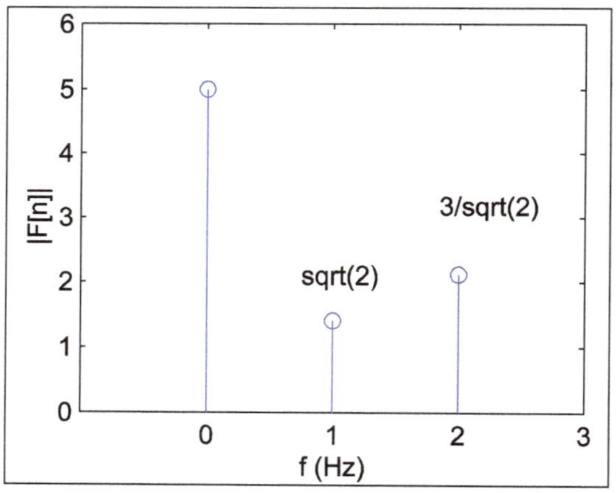

DFT of four point signal.

Thus, the conventional way of displaying a spectrum is not as shown in figure: In typical applications, N is much greater than Œ; for example, for $N=1024$, $F[n]$ has 1024 components, but $513-1023$ are the complex conjugates of $511-1$, leaving $\frac{F[0]}{1024}$ as the d.c. component, $\frac{2}{1024}\frac{|F[1]|}{\sqrt{2}}$ to $\frac{2}{1024}\frac{|F[511]|}{\sqrt{2}}$ as complete a.c. components and $\frac{1}{1024}\frac{F[512]}{\sqrt{2}}$ as the cosine-only component at the highest distinguishable frequency $\left(n=\frac{N}{2}\right)$.

Most computer programmes evaluate $\frac{|F[n]|}{N}$ (or $\frac{|F[n]|^2}{N}$: for the power spectral density) which gives the correct "shape" for the spectrum, except for the values at $n = 0$ and $\frac{N}{2}$.

Discrete Fourier Transform Errors

To what degree does the DFT approximate the Fourier transform of the function underlying the data? Clearly the DFT is only an approximation since it provides only for a finite set of frequencies. But how correct are these discrete values themselves? There are two main types of DFT errors: aliasing and "leakage":

Aliasing

This is another manifestation of the phenomenon which we have now encountered several times. If the initial samples are not sufficiently closely spaced to represent high-frequency components present in the underlying function, then the DFT values will be corrupted by aliasing. As before, the solution is either to increase the sampling rate (if possible) or to pre-filter the signal in order to minimise its highfrequency spectral content.

Leakage

Recall that the continuous Fourier transform of a periodic waveform requires the integration to be performed over the $-\infty$ to $+\infty$ - or over an integer number of cycles of the waveform. If we attempt to complete the DFT over a non-integer number of cycles of the input signal, then we might expect the transform to be corrupted in some way. This is indeed the case, as will now be shown.

Consider the case of an input signal which is a sinusoid with a fractional number of cycles in the N data samples. The DFT for this case $\left(for\ n = 0\ to\ n = \frac{N}{2} \right)$.

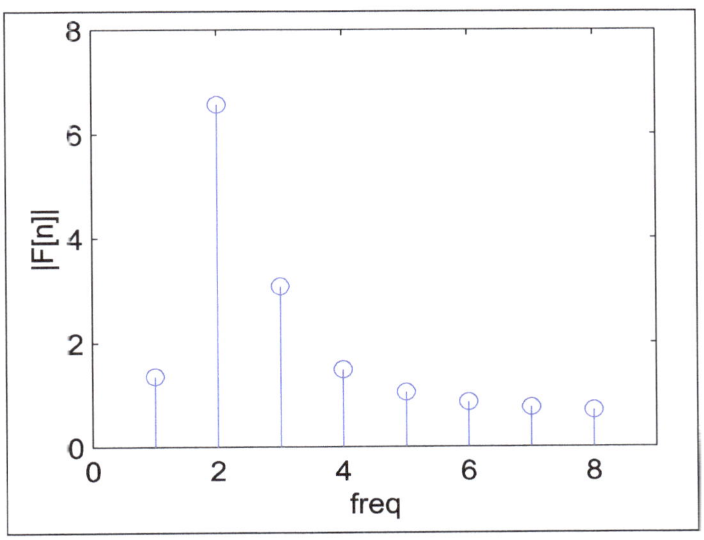

Leakage.

We might have expected the DFT to give an output at just the quantised frequencies either side of the true frequency. This certainly does happen but we also find non-zero outputs at all other frequencies. This smearing effect, which is known as leakage, arises because we are effectively calculating the Fourier series for the waveform in figure, which has major discontinuities, hence other frequency components.

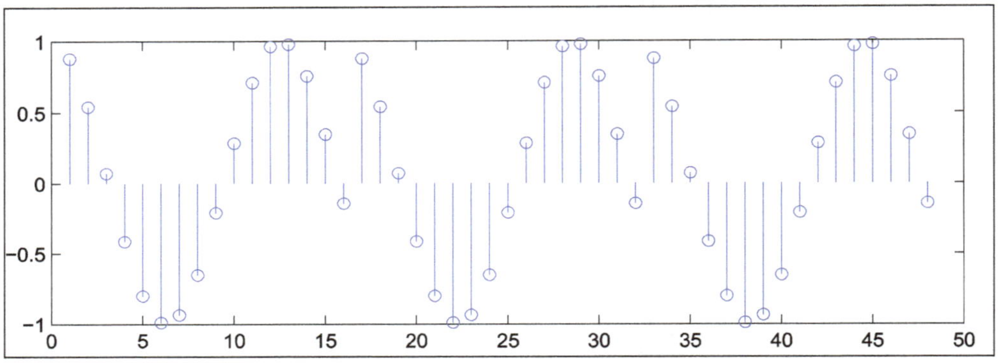

Leakage: The repeating waveform has discontinuities.

Most sequences of real data are much more complicated than the sinusoidal sequences that we have so far considered and so it will not be possible to avoid introducing discontinuities when using a finite number of points from the sequence in order to calculate the DFT.

The solution is to use one of the window functions which we encountered in the design of FIR filters (e.g. the Hamming or Hanning windows). These window functions taper the samples towards zero values at both endpoints, and so there is no discontinuity (or very little, in the case of the Hanning window) with a hypothetical next period. Hence the leakage of spectral content away from its correct location is much reduced, as in figure.

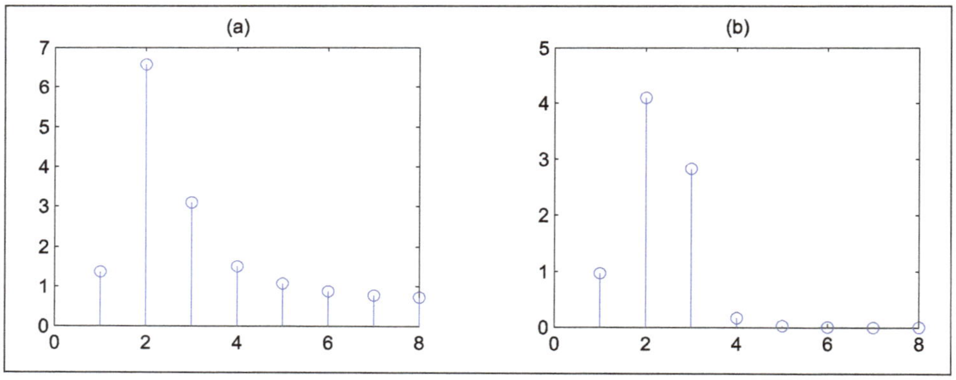

Leakage is reduced using a Hanning window.

The Fast Fourier Transform

he time taken to evaluate a DFT on a digital computer depends principally on the number of multiplications involved, since these are the slowest operations. With the DFT, this number is directly related to V (matrix multiplication of a vector), where is the length of the transform. For most problems, is chosen to be at least 256 in order to get a reasonable approximation for the spectrum of the sequence under consideration – hence computational speed becomes a major consideration.

Highly efficient computer algorithms for estimating Discrete Fourier Transforms have been developed since the mid-60's. These are known as Fast Fourier Transform (FFT) algorithms and they rely on the fact that the standard DFT involves a lot of redundant calculations:

Re-writing $F[n]=\sum_{k=0}^{N-1} f[k]e^{-j\frac{2\pi}{N}nk}$ as $F[n]=\sum_{k=0}^{N-1} f[k]W_N^{nk}$

it is easy to realise that the same values of W_N^{nk} are calculated many times as the computation proceeds. Firstly, the integer product nk repeats for different combinations of k and n ; secondly, W_N^{nk} is a periodic function with only N distinct values.

For example, cons $N=8$ &9 (the FFT is simplest by far if is N an integral power of 2)

$$W_8^1=e^{-j\frac{2\pi}{8}}=e^{-j45^\circ}=\frac{1-j}{\sqrt{2}}=a, \text{ say.}$$

Then $a^2=-j \quad a^3=-ja=-a* \quad a^4=-1$

$$a^5=-a \quad a^6=-j \quad a^7=ja-a* \quad a^8=1$$

From the above, it can be seen that:

$$W_8^4=-W^0$$
$$W_8^5=-W_1^8$$
$$W_8^6=-W_2^8$$
$$W_8^7=-W_8^3$$

Also, if nk falls outside the range 0-7, we still get one of the above values:

eg. If $n=5$ and $k=7$, $W_8^{35}=a^{35}=(a^8)^4\cdot a^3=a^3$

Decimation-in-time Algorithm

Let us begin by splitting the single summation over N samples into 2 summations, each with $\frac{N}{2}$ samples, one for k even and the other for k odd.

Substitute $m=\frac{k}{2}$ for k even and $m=\frac{k-1}{2}$ for k odd and write:

$$F[n]=\sum_{m=0}^{\frac{N}{2}-1} f[2m]W_N^{2mn}+\sum_{m=0}^{\frac{N}{2}-1} f[2m+1]W_N^{(2m+1)n}$$

Note that $W_N^{2mn}=e^{-j\frac{2\pi}{N}(2mn)}=e^{-j\frac{2\pi}{\frac{N}{2}}mn}=W_{\frac{N}{2}}^{mn}$

Therefore $F[n]=\sum_{m=0}^{\frac{N}{2}-1} f[2m]W_{\frac{N}{2}}^{mn}+W_N^m\sum_{m=0}^{\frac{N}{2}-1} f[2m+1]W_{\frac{N}{2}}^{mn}$

ie. $F[n] = G[n] + W_N^m H[n]$

Thus the N-point DET $F[n]$ can be obtained from two $\frac{N}{2}$–point transforms, one on even input data, $G[n]$, and one on odd input data, $H[n]$. Although the frequency index n ranges over N values, only $\frac{N}{2}$ values of $G[n]$ and $H[n]$ need to be computed since $G[n]$ and $H[n]$ are periodic in n with period $\frac{N}{2}$.

For example, for $N = 8$:

- Even input data $f[0]\, f[2]\, f[4]\, f[6]$
- Odd input data $f[1]\, f[3]\, f[5]\, f[7]$

$$F[0] = G[0] + W_8^0 H[0]$$
$$F[1] = G[1] + W_8^1 H[1]$$
$$F[2] = G[2] + W_8^2 H[2]$$
$$F[3] = G[3] + W_8^3 H[3]$$
$$F[4] = G[0] + W_8^4 H[0] = G[0] - W_8^0 H[0]$$
$$F[5] = G[1] + W_8^5 H[1] = G[1] - W_8^1 H[1]$$
$$F[6] = G[2] + W_8^6 H[2] = G[2] - W_8^2 H[2]$$
$$F[7] = G[3] + W_8^7 H[3] = G[3] - W_8^3 H[3]$$

This is shown graphically on the flow graph of figure:

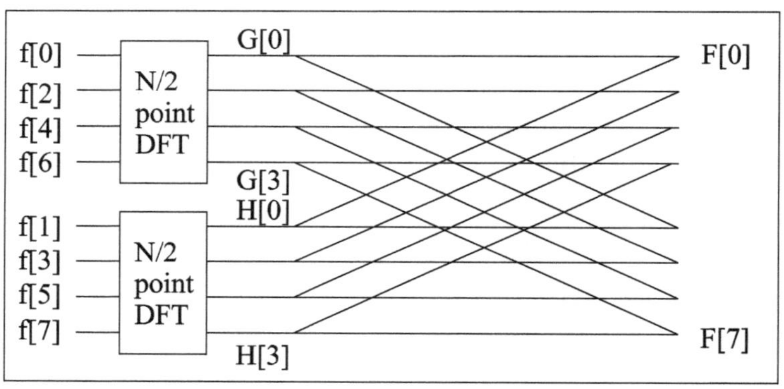

FFT flow graph.

Assuming than N is a power of 2, we can repeat the above process on the two $\frac{N}{2}$–-point transforms, breaking them down to $\frac{N}{4}$– point transforms, etc., until we come down to 2-point transforms. For $N = 8$, , only one further stage is needed (i.e. there are ã $N = 2^{ã}$), as shown below in figure.

Thus the FFT is computed by dividing up, or decimating, the sample sequence $f[k]$ into sub-sequences until only 2-point DFT's remain. Since it is the input, or time, samples which are divided up, this algorithm is known as the decimationin-time (DIT) algorithm. (An equivalent algorithm exists for which the output, or frequency, points are sub-divided – the decimation-in-frequency algorithm.)

The basic computation at the heart of the FFT is known as the butterfly because of its criss-cross appearance. For the DIT FFT algorithm, the butterfly computation is of the form of figure.

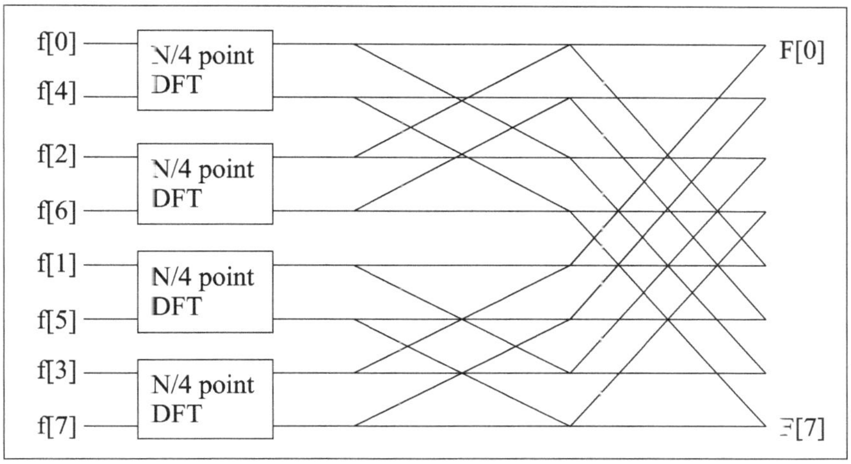

FFT flow graph.

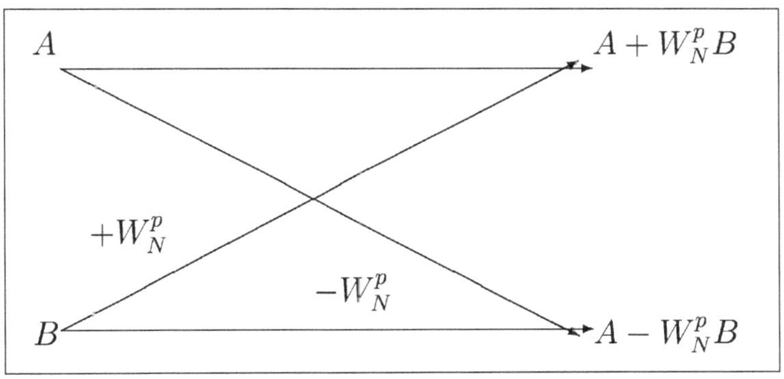

Butterfly operation in FFT.

where A and B are complex numbers. Thus a butterfly computation requires one complex multiplication and 2 complex additions.

Note also, that the input samples are "bit-reversed" because at each stage of decimation the sequence input samples is separated into even- and odd- indexed samples.

(The bit-reversal algorithm only applies if N is an integral power of 2).

Index (k)	Binary Representation	Bit-reversed Binary	Bit-reversed Index
0	000	000	0
1	001	100	4

2	010	010	2
3	011	110	6
4	100	001	1
5	101	101	5
6	110	011	3
7	111	111	7

Computational Speed of FFT

The DFT requires N^2 complex multiplications. At each stage of the FFT (i.e. each halving) $\frac{N}{2}$ complex multiplications are required to combine the results of the previous stage. Since there are $(\log_2 N)$ stages, the number of complex multiplications required to evaluate an -point DFT with the FFT is approximately $N/2\log_2 N$ (approximately because multiplications by factors such as $W_N^0, W_N^{\frac{N}{2}}, W_N^{\frac{N}{4}}$ *and* $W_N^{\frac{3N}{4}}$ are really just complex additions and subtractions.

N	N^2(DFT)	$\frac{N}{2}\log_2 N$(FFT)	saving
32	1,024	80	92%
256	65,536	1,024	98%
1,024	1,048,576	5,120	99.5%

Practical Considerations

If is not a power of 2, there are 2 strategies available to complete an N -point FFT.

- Take advantage of such factors as N possesses. For example, if N is divisible by 3 (e.g. N =48), the final decimation stage would include a 3-point transform.
- Pack the data with zeroes; e.g. include 16 zeroes with the 48 data points (for N =48) and compute 64-point FFT. (However, you should again be wary of abrupt transitions between the trailing (or leading) edge of the data and the following (or preceding) zeroes; a better approach might be to pack the data with more realistic "dummy values").

Fast Fourier Transform

A fast Fourier transform (FFT) is an algorithm that computes the discrete Fourier transform (DFT) of a sequence, or its inverse (IDFT). Fourier analysis converts a signal from its original domain (often time or space) to a representation in the frequency domain and vice versa. The DFT is obtained by decomposing a sequence of values into components of different frequencies. This operation is useful in many fields, but computing it directly from the definition is often too slow to be practical. An FFT rapidly computes such transformations by factorizing the DFT matrix into a product of sparse (mostly zero) factors. As a result, it manages to reduce the complexity of computing the DFT from $O(N^2)$, which arises if one simply applies the definition of DFT, to $O(N \log N)$, where N is the data size. The difference in speed can be enormous, especially for long data sets where N may be in the thousands or millions. In the presence of round-off error, many FFT algorithms are much more accurate than evaluating the DFT definition directly or indirectly. There are many different FFT algorithms based on a wide range of published theories, from simple complex-number arithmetic to group theory and number theory.

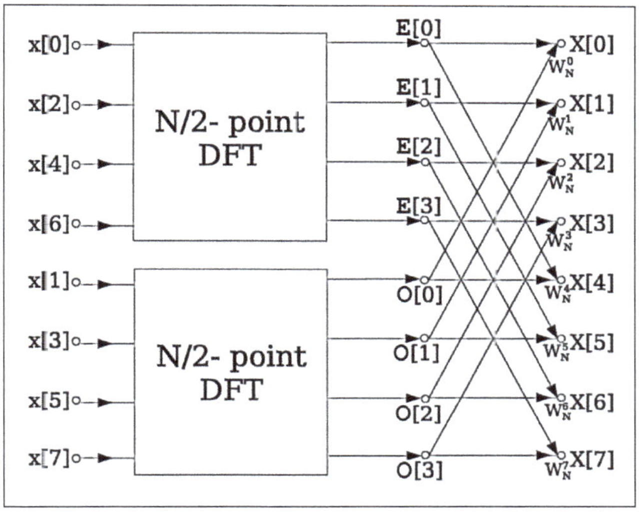

An example FFT algorithm structure, using a decomposition into half-size FFTs.

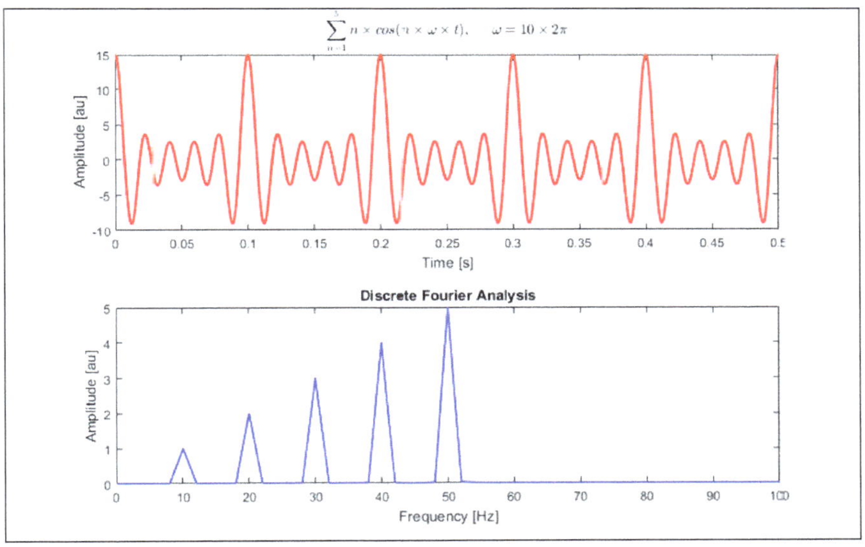

A discrete Fourier analysis of a sum of cosine waves at 10, 20, 30, 40, and 50 Hz.

The best-known FFT algorithms depend upon the factorization of N, but there are FFTs with $O(N \log N)$ complexity for all N, even for prime N. Many FFT algorithms only depend on the fact that $e^{-2\pi i/N}$ is an N-th primitive root of unity, and thus can be applied to analogous transforms over any finite field, such as number-theoretic transforms. Since the inverse DFT is the same as the

DFT, but with the opposite sign in the exponent and a $1/N$ factor, any FFT algorithm can easily be adapted for it.

Let x_0, ..., x_{N-1} be complex numbers. The DFT is defined by the formula,

$$X_k = \sum_{n=0}^{N-1} x_n e^{-i2\pi kn/N} \qquad k = 0,\ldots,N-1,$$

where $e^{i2\pi/N}$ is a primitive N the root of 1.

Evaluating this definition directly requires $O(N^2)$ operations: there are N outputs X_k, and each output requires a sum of N terms. An FFT is any method to compute the same results in operations. All known FFT algorithms require $\Theta(N \log N)$ operations, although there is no known proof that a lower complexity score is impossible.

To illustrate the savings of an FFT, consider the count of complex multiplications and additions for N=4096 data points. Evaluating the DFT's sums directly involves N^2 complex multiplications and $N(N-1)$ complex additions, of which $O(N)$ operations can be saved by eliminating trivial operations such as multiplications by 1, leaving about 30 million operations. On the other hand, the radix-2 Cooley–Tukey algorithm, for N a power of 2, can compute the same result with only $(N/2)\log_2(N)$ complex multiplications (again, ignoring simplifications of multiplications by 1 and similar) and $N \log_2(N)$ complex additions, in total about 30,000 operations - a thousand times less than with direct evaluation. In practice, actual performance on modern computers is usually dominated by factors other than the speed of arithmetic operations and the analysis is a complicated subject, but the overall improvement from $O(N^2)$ to $O(N \log N)$ remains.

Algorithms

Cooley–tukey Algorithm

By far the most commonly used FFT is the Cooley–Tukey algorithm. This is a divide and conquer algorithm that recursively breaks down a DFT of any composite size $N = N_1 N_2$ into many smaller DFTs of sizes N_1 and N_2, along with O(N) multiplications by complex roots of unity traditionally called twiddle factors.

This method (and the general idea of an FFT) was popularized in 1965, but it was later discovered that those two authors had independently re-invented an algorithm known to Carl Friedrich Gauss around 1805 (and subsequently rediscovered several times in limited forms).

The best known use of the Cooley–Tukey algorithm is to divide the transform into two pieces of size $N/2$ at each step, and is therefore limited to power-of-two sizes, but any factorization can be used in general (as was known to both Gauss and Cooley/Tukey). These are called the *radix-2* and *mixed-radix* cases, respectively (and other variants such as the split-radix FFT have their own names as well). Although the basic idea is recursive, most traditional implementations rearrange the algorithm to avoid explicit recursion. Also, because the Cooley–Tukey algorithm breaks the DFT into smaller DFTs, it can be combined arbitrarily with any other algorithm for the DFT, such as those described below.

Other FFT Algorithms

There are other FFT algorithms distinct from Cooley–Tukey.

Cornelius Lanczos did pioneering work on the FFT and FFS (fast Fourier sampling method) with G. C. Danielson.

For $N = N_1N_2$ with coprime N_1 and N_2, one can use the prime-factor (Good–Thomas) algorithm (PFA), based on the Chinese remainder theorem, to factorize the DFT similarly to Cooley–Tukey but without the twiddle factors. The Rader–Brenner algorithm is a Cooley–Tukey-like factorization but with purely imaginary twiddle factors, reducing multiplications at the cost of increased additions and reduced numerical stability; it was later superseded by the split-radix variant of Cooley–Tukey (which achieves the same multiplication count but with fewer additions and without sacrificing accuracy). Algorithms that recursively factorize the DFT into smaller operations other than DFTs include the Bruun and QFT algorithms. (The Rader–Brenner and QFT algorithms were proposed for power-of-two sizes, but it is possible that they could be adapted to general composite N. Bruun's algorithm applies to arbitrary even composite sizes.) Bruun's algorithm, in particular, is based on interpreting the FFT as a recursive factorization of the polynomial $z^N - 1$, here into real-coefficient polynomials of the form $z^M - 1$ and $z^{2M} + az^M + 1$.

Another polynomial viewpoint is exploited by the Winograd FFT algorithm, which factorizes $z^N - 1$ into cyclotomic polynomials—these often have coefficients of 1, 0, or −1, and therefore require few (if any) multiplications, so Winograd can be used to obtain minimal-multiplication FFTs and is often used to find efficient algorithms for small factors. Indeed, Winograd showed that the DFT can be computed with only $O(N)$ irrational multiplications, leading to a proven achievable lower bound on the number of multiplications for power-of-two sizes; unfortunately, this comes at the cost of many more additions, a tradeoff no longer favorable on modern processors with hardware multipliers. In particular, Winograd also makes use of the PFA as well as an algorithm by Rader for FFTs of *prime* sizes.

Rader's algorithm, exploiting the existence of a generator for the multiplicative group modulo prime N, expresses a DFT of prime size N as a cyclic convolution of (composite) size N–1, which can then be computed by a pair of ordinary FFTs via the convolution theorem (although Winograd uses other convolution methods). Another prime-size FFT is due to L. I. Bluestein, and is sometimes called the chirp-z algorithm; it also re-expresses a DFT as a convolution, but this time of the *same* size (which can be zero-padded to a power of two and evaluated by radix-2 Cooley–Tukey FFTs, for example), via the identity

$$nk = -\frac{(k-n)^2}{2} + \frac{n^2}{2} + \frac{k^2}{2}.$$

Hexagonal Fast Fourier Transform aims at computing an efficient FFT for the hexagonally sampled data by using a new addressing scheme for hexagonal grids, called Array Set Addressing (ASA).

FFT Algorithms Specialized for Real and Symmetric Data

In many applications, the input data for the DFT are purely real, in which case the outputs satisfy the symmetry:

$$X_{N-k} = X_k^*$$

and efficient FFT algorithms have been designed for this situation. One approach consists of taking an ordinary algorithm (e.g. Cooley–Tukey) and removing the redundant parts of the computation, saving roughly a factor of two in time and memory. Alternatively, it is possible to express an *even*-length real-input DFT as a complex DFT of half the length (whose real and imaginary parts are the even/odd elements of the original real data), followed by O(*N*) post-processing operations.

It was once believed that real-input DFTs could be more efficiently computed by means of the discrete Hartley transform (DHT), but it was subsequently argued that a specialized real-input DFT algorithm (FFT) can typically be found that requires fewer operations than the corresponding DHT algorithm (FHT) for the same number of inputs. Bruun's algorithm (above) is another method that was initially proposed to take advantage of real inputs, but it has not proved popular.

There are further FFT specializations for the cases of real data that have even/odd symmetry, in which case one can gain another factor of (roughly) two in time and memory and the DFT becomes the discrete cosine/sine transform(s) (DCT/DST). Instead of directly modifying an FFT algorithm for these cases, DCTs/DSTs can also be computed via FFTs of real data combined with O(*N*) pre/post processing.

Computational Issues

Bounds on Complexity and Operation Counts

A fundamental question of longstanding theoretical interest is to prove lower bounds on the complexity and exact operation counts of fast Fourier transforms, and many open problems remain. It is not even rigorously proved whether DFTs truly require $\Omega(N \log N)$ (i.e., order $N \log N$ or greater) operations, even for the simple case of power of two sizes, although no algorithms with lower complexity are known. In particular, the count of arithmetic operations is usually the focus of such questions, although actual performance on modern-day computers is determined by many other factors such as cache or CPU pipeline optimization.

Following pioneering work by Winograd, a tight $\Theta(N)$ lower bound *is* known for the number of real multiplications required by an FFT. It can be shown that only $4N - 2\log_2^2 N - 2\log_2 N - 4$ irrational real multiplications are required to compute a DFT of power-of-two length $N = 2^m$. Moreover, explicit algorithms that achieve this count are known. Unfortunately, these algorithms require too many additions to be practical, at least on modern computers with hardware multipliers.

A tight lower bound is *not* known on the number of required additions, although lower bounds have been proved under some restrictive assumptions on the algorithms. In 1973, Morgenstern proved an $\Omega(N \log N)$ lower bound on the addition count for algorithms where the multiplicative constants have bounded magnitudes (which is true for most but not all FFT algorithms). This result, however, applies only to the unnormalized Fourier transform (which is a scaling of a unitary matrix by a factor of $\sqrt{N}$), and does not explain why the Fourier matrix is harder to compute than any other unitary matrix (including the identity matrix) under the same scaling. Pan proved an $\Omega(N \log N)$ lower bound assuming a bound on a measure of the FFT algorithm's "asynchronicity", but the generality of this assumption is unclear. For the case of power-of-two *N*, Papadimitriou argued that the number $N \log_2 N$ of complex-number additions achieved by Cooley–Tukey algorithms is *optimal* under certain assumptions on the graph of the algorithm (his assumptions imply, among other things, that no additive identities in the roots of unity are exploited). (This argument would imply that at least $2N \log_2 N$ real additions are required, although this is not a tight bound because extra additions are required as part of complex-number multiplications.) Thus far, no published FFT algorithm has achieved fewer than $N \log_2 N$ complex-number additions (or their equivalent) for power-of-two *N*.

A third problem is to minimize the *total* number of real multiplications and additions, sometimes called the "arithmetic complexity" (although in this context it is the exact count and not the asymptotic complexity that is being considered). Again, no tight lower bound has been proven. Since 1968, however, the lowest published count for power-of-two *N* was long achieved by the split-radix FFT algorithm, which requires $4N \log_2 N - 6N + 8$ real multiplications and additions for $N >$ This was recently reduced to $\sim \frac{34}{9} N \log_2 N$ (Johnson and Frigo, A slightly larger count (but still better than split radix for *N*≥256) was shown to be provably optimal for *N*≤512 under additional restrictions on the possible algorithms (split-radix-like flowgraphs with unit-modulus multiplicative factors), by reduction to a satisfiability modulo theories problem solvable by brute force.

Most of the attempts to lower or prove the complexity of FFT algorithms have focused on the ordinary complex-data case, because it is the simplest. However, complex-data FFTs are so closely related to algorithms for related problems such as real-data FFTs, discrete cosine transforms, discrete Hartley transforms, and so on, that any improvement in one of these would immediately lead to improvements in the others.

Approximations

All of the FFT algorithms discussed above compute the DFT exactly (i.e. neglecting floating-point errors). A few "FFT" algorithms have been proposed, however, that compute the DFT *approximately*, with an error that can be made arbitrarily small at the expense of increased computations. Such algorithms trade the approximation error for increased speed or other properties. For example, an approximate FFT algorithm by Edelman et al. achieves lower communication requirements for parallel computing with the help of a fast multipole method. A wavelet-based approximate FFT by Guo and Burrus takes sparse inputs/outputs (time/frequency localization) into account more efficiently than is possible with an exact FFT. Another algorithm for approximate computation of a subset of the DFT outputs is due to Shentov et al. The Edelman algorithm works equally well for sparse and non-sparse data, since it is based on the compressibility (rank deficiency) of the Fourier matrix itself rather than the compressibility (sparsity) of the data. Conversely, if the data are sparse—that is, if only *K* out of *N* Fourier coefficients are nonzero—then the complexity can be reduced to $O(K \log(N) \log(N/K))$, and this has been demonstrated to lead to practical speedups compared to an ordinary FFT for $N/K > 32$ in a large-*N* example ($N = 2^{22}$) using a probabilistic approximate algorithm (which estimates the largest *K* coefficients to several decimal places).

Accuracy

Even the "exact" FFT algorithms have errors when finite-precision floating-point arithmetic is used, but these errors are typically quite small; most FFT algorithms, e.g. Cooley–Tukey, have excellent numerical properties as a consequence of the pairwise summation structure of the algorithms. The upper bound on the relative error for the Cooley–Tukey algorithm is O(ε log N), compared to O($\varepsilon N^{3/2}$) for the naïve DFT formula, where ε is the machine floating-point relative precision. In fact, the root mean square (rms) errors are much better than these upper bounds, being only O(ε $\sqrt{\log N}$) for Cooley–Tukey and O(ε $\sqrt{N}$) for the naïve DFT. These results, however, are very sensitive to the accuracy of the twiddle factors used in the FFT (i.e. the trigonometric function values), and it is not unusual for incautious FFT implementations to have much worse accuracy, e.g. if they use inaccurate trigonometric recurrence formulas. Some FFTs other than Cooley–Tukey, such as the Rader–Brenner algorithm, are intrinsically less stable.

In fixed-point arithmetic, the finite-precision errors accumulated by FFT algorithms are worse, with rms errors growing as O($\sqrt{N}$) for the Cooley–Tukey algorithm. Moreover, even achieving this accuracy requires careful attention to scaling to minimize loss of precision, and fixed-point FFT algorithms involve rescaling at each intermediate stage of decompositions like Cooley–Tukey.

To verify the correctness of an FFT implementation, rigorous guarantees can be obtained in O(N log N) time by a simple procedure checking the linearity, impulse-response, and time-shift properties of the transform on random inputs.

Multidimensional FFTs

The multidimensional DFT

$$X_{\mathbf{k}} = \sum_{\mathbf{n}=0}^{\mathbf{N}-1} e^{-2\pi i \mathbf{k}\cdot(\mathbf{n}/\mathbf{N})} x_{\mathbf{n}}$$

transforms an array $x_{\mathbf{n}}$ with a d-dimensional vector of indices $\mathrm{n} = (n_1, \ldots, n_d)$ by a set of d nested summations (over $n_j = 0 \ldots N_j - 1$ for each j), where the division n/N, defined as $\mathrm{n}/\mathrm{N} = (n_1/N_1, \ldots, n_d/N_d)$, is performed element-wise. Equivalently, it is the composition of a sequence of d sets of one-dimensional DFTs, performed along one dimension at a time (in any order).

This compositional viewpoint immediately provides the simplest and most common multidimensional DFT algorithm, known as the row-column algorithm. That is, one simply performs a sequence of d one-dimensional FFTs (by any of the above algorithms): first you transform along the n_1 dimension, then along the n_2 dimension, and so on (or actually, any ordering works). This method is easily shown to have the usual O(N log N) complexity, where $N = N_1 \cdot N_2 \cdot \ldots \cdot N_d$ is the total number of data points transformed. In particular, there are N/N_1 transforms of size N_1, etcetera, so the complexity of the sequence of FFTs is:

$$\frac{N}{N_1} O(N_1 \log N_1) + \cdots + \frac{N}{N_d} O(N_d \log N_d)$$
$$= O\left(N\left[\log N_1 + \cdots + \log N_d\right]\right) = O(N \log N).$$

In two dimensions, the $x_{\mathbf{k}}$ can be viewed as an $n_1 \times n_2$ matrix, and this algorithm corresponds to first performing the FFT of all the rows (resp. columns), grouping the resulting transformed rows (resp. columns) together as another $n_1 \times n_2$ matrix, and then performing the FFT on each of the columns (resp. rows) of this second matrix, and similarly grouping the results into the final result matrix.

In more than two dimensions, it is often advantageous for cache locality to group the dimensions recursively. For example, a three-dimensional FFT might first perform two-dimensional FFTs of each planar "slice" for each fixed n_1, and then perform the one-dimensional FFTs along the n_1 direction. More generally, an asymptotically optimal cache-oblivious algorithm consists of recursively dividing the dimensions into two groups $(n_1, \ldots, n_{d/2})$ and $(n_{d/2+1}, \ldots, n_d)$ that are transformed recursively (rounding if d is not even) .Still, this remains a straightforward variation of the row-column algorithm that ultimately requires only a one-dimensional FFT algorithm as the base case, and still has O(N log N) complexity. Yet another variation is to perform matrix transpositions in between transforming subsequent dimensions, so that the transforms operate on contiguous data; this is especially important for out-of-core and distributed memory situations where accessing non-contiguous data is extremely time-consuming.

There are other multidimensional FFT algorithms that are distinct from the row-column algorithm, although all of them have O(N log N) complexity. Perhaps the simplest non-row-column FFT is the vector-radix FFT algorithm, which is a generalization of the ordinary Cooley–Tukey algorithm where one divides the transform dimensions by a vector $\mathrm{r} = (r_1, r_2, \ldots, r_d)$ of radices at each step. (This may also have cache benefits.) The simplest case of vector-radix is where all of the radices are equal (e.g. vector-radix-2 divides *all* of the dimensions by two), but this is not necessary. Vector radix with only a single non-unit radix at a time, i.e. $\mathrm{r} = (1, \ldots, 1, r, 1, \ldots, 1)$, is essentially a row-column algorithm. Other, more complicated, methods include polynomial transform algorithms due to Nussbaumer, which view the transform in terms of convolutions and polynomial products.

Other Generalizations

An O($N^{5/2}$log N) generalization to spherical harmonics on the sphere S^2 with N^2 nodes was described by Mohlenkamp, along with an algorithm conjectured (but not proven) to have O(N^2 log^2(N)) complexity; Mohlenkamp also provides an implementation in the libftsh library. A spherical-harmonic algorithm with O(N^2log N) complexity is described by Rokhlin and Tygert.

The fast folding algorithm is analogous to the FFT, except that it operates on a series of binned waveforms rather than a series of real or complex scalar values. Rotation (which in the FFT is multiplication by a complex phasor) is a circular shift of the component waveform.

Various groups have also published "FFT" algorithms for non-equispaced data, as reviewed in Potts *et al.* Such algorithms do not strictly compute the DFT (which is only defined for equispaced data), but rather some approximation thereof (a non-uniform discrete Fourier transform, or NDFT, which itself is often computed only approximately). More generally there are various other methods of spectral estimation.

Applications

The FFT is used in digital recording, sampling, additive synthesis and pitch correction software.

The FFT's importance derives from the fact that it has made working in the frequency domain equally computationally feasible as working in the temporal or spatial domain. Some of the important applications of the FFT include:

- Fast large-integer and polynomial multiplication.
- Efficient matrix-vector multiplication for Toeplitz, circulant and other structured matrices.
- Filtering algorithms.
- Fast algorithms for discrete cosine or sine transforms (e.g., fast DCT used for JPEG and MP_3/MPEG encoding/decoding).
- Fast Chebyshev approximation.
- Solving difference equations.
- Computation of isotopic distributions.

Big FFTs: With the explosion of big data in fields such as astronomy, the need for 512k FFTs has arisen for certain interferometry calculations. The data collected by projects such as WMAP and LIGO require FFTs of tens of billions of points. As this size does not fit into main memory, so called out-of-core FFTs are an active area of research.

Approximate FFTs: For applications such as MRI, it is necessary to compute DFTs for nonuniformly spaced grid points and frequencies. Multipole based approaches can compute approximate quantities with factor of runtime increase.

Group FFTs: The FFT may also be explained and interpreted using group representation theory that allows for further generalization. A function on any compact group, including non cyclic, has an expansion in terms of a basis of irreducible matrix elements. It remains active area of research to find efficient algorithm for performing this change of basis. Applications including efficient spherical harmonic expansion, analyzing certain Markov processes, robotics etc.

Quantum FFTs: Shor's fast algorithm for integer factorization on a quantum computer has a subroutine to compute DFT of a binary vector. This is implemented as sequence of 1- or 2-bit quantum gates now known as quantum FFT, which is effectively the Cooley–Tukey FFT realized as a particular factorization of the Fourier matrix. Extension to these ideas is currently being explored.

Spectral Density

The power spectrum $S_{xx}(f$ of a time series $x(t)$ describes the distribution of power into frequency components composing that signal. According to Fourier analysis, any physical signal can be decomposed into a number of discrete frequencies, or a spectrum of frequencies over a continuous range. The statistical average of a certain signal or sort of signal (including noise) as analyzed in terms of its frequency content, is called its spectrum.

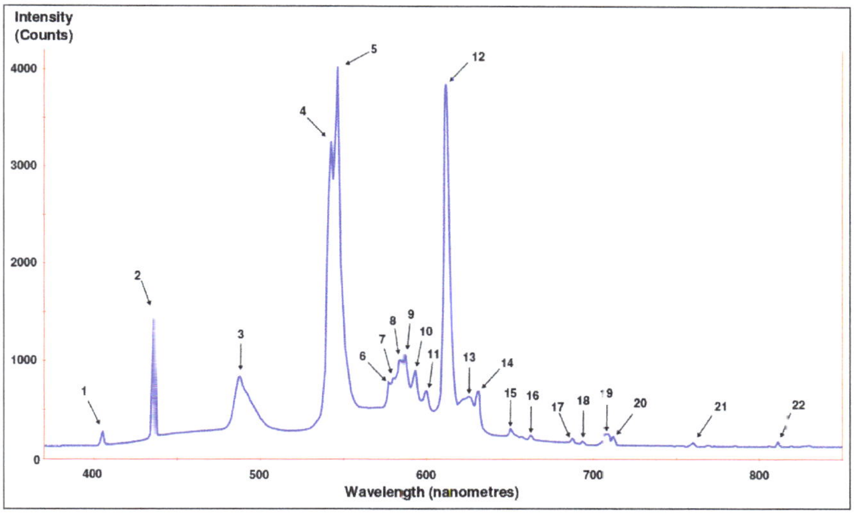

The spectral density of a fluorescent light as a function of optical wavelength shows peaks at atomic transitions, indicated by the numbered arrows.

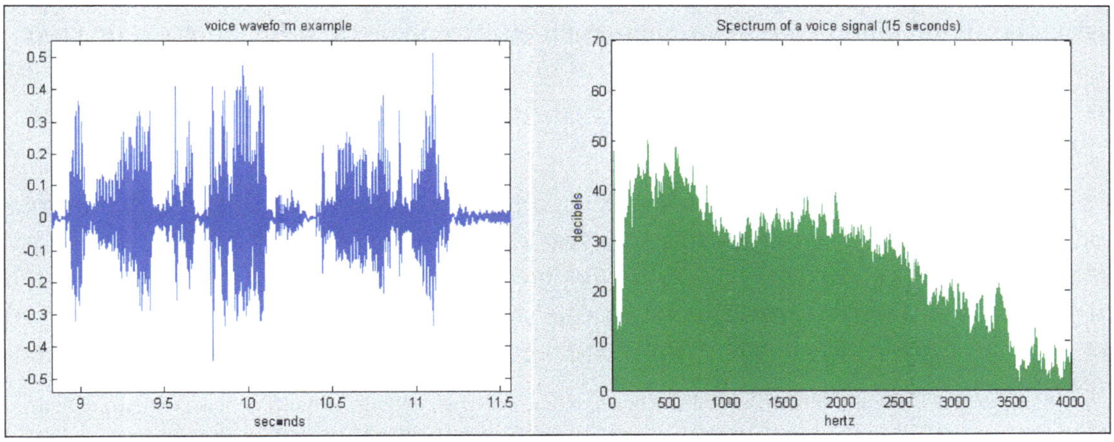

The voice waveform over time (left) has a broad audio power spectrum (right).

When the energy of the signal is concentrated around a finite time interval, especially if its total energy is finite, one may compute the energy spectral density. More commonly used is the power spectral density (or simply power spectrum), which applies to signals existing over *all* time, or over a time period large enough (especially in relation to the duration of a measurement) that it could as well have been over an infinite time interval. The power spectral density (PSD) then refers to the spectral energy distribution that would be found per unit time, since the total energy of such a signal over all time would generally be infinite. Summation or integration of the spectral components yields the total power (for a physical process) or variance (in a statistical process), identical to what would be obtained by integrating $x^2(t)$ over the time domain, as dictated by Parseval's theorem.

The spectrum of a physical process $x(t)$ often contains essential information about the nature of x. For instance, the pitch and timbre of a musical instrument are immediately determined from a spectral analysis. The color of a light source is determined by the spectrum of the electromagnetic wave's electric field $E(t)$ as it fluctuates at an extremely high frequency. Obtaining a spectrum from time series such as these involves the Fourier transform, and generalizations based on Fourier analysis. In many cases the time domain is not specifically employed in practice, such as when a dispersive prism is used to obtain a spectrum of light in a spectrograph, or when a sound is perceived through its effect on the auditory receptors of the inner ear, each of which is sensitive to a particular frequency.

Any signal that can be represented as a variable that varies in time has a corresponding frequency spectrum. This includes familiar entities such as visible light (perceived as color), musical notes (perceived as pitch), radio/TV (specified by their frequency, or sometimes wavelength) and even the regular rotation of the earth. When these signals are viewed in the form of a frequency spectrum, certain aspects of the received signals or the underlying processes producing them are revealed. In some cases the frequency spectrum may include a distinct peak corresponding to a sine wave component. And additionally there may be peaks corresponding to harmonics of a fundamental peak, indicating a periodic signal which is *not* simply sinusoidal. Or a continuous spectrum may show narrow frequency intervals which are strongly enhanced corresponding to resonances, or frequency intervals containing almost zero power as would be produced by a notch filter.

In physics, the signal might be a wave, such as an electromagnetic wave, an acoustic wave, or the vibration of a mechanism. The *power spectral density* (PSD) of the signal describes the power present in the signal as a function of frequency, per unit frequency. Power spectral density is commonly expressed in watts per hertz (W/Hz).

When a signal is defined in terms only of a voltage, for instance, there is no unique power associated with the stated amplitude. In this case "power" is simply reckoned in terms of the square of the signal, as this would always be *proportional* to the actual power delivered by that signal into a given impedance. So one might use units of $V^2\ Hz^{-1}$ for the PSD and $V^2\ s\ Hz^{-1}$ for the ESD (*energy spectral density*) even though no actual "power" or "energy" is specified.

Sometimes one encounters an *amplitude spectral density* (ASD), which is the square root of the PSD; the ASD of a voltage signal has units of $V\ Hz^{-1/2}$. This is useful when the *shape* of the spectrum is rather constant, since variations in the ASD will then be proportional to variations in the signal's voltage level itself. But it is mathematically preferred to use the PSD, since only in that case is the area under the curve meaningful in terms of actual power over all frequency or over a specified bandwidth.

In the general case, the units of PSD will be the ratio of units of variance per unit of frequency; so, for example, a series of displacement values (in meters) over time (in seconds) will have PSD in units of m^2/Hz. For random vibration analysis, units of $g^2\ Hz^{-1}$ are frequently used for the PSD of acceleration. Here g denotes the g-force.

Mathematically, it is not necessary to assign physical dimensions to the signal or to the independent variable.

Energy Spectral Density

Energy spectral density describes how the energy of a signal or a time series is distributed with frequency. Here, the term energy is used in the generalized sense of signal processing; that is, the energy E of a signal $x(t)$ is,

$$E = \int_{-\infty}^{\infty} | x(t) |^2 dt.$$

The energy spectral density is most suitable for transients—that is, pulse-like signals—having a finite total energy. In this case, Parseval's theorem gives us an alternate expression for the energy of the signal:

$$\hat{x}(f) = \int_{-\infty}^{\infty} | x(t) |^2 dt = \int_{-\infty}^{\infty} | \hat{x}(f) |^2 df,$$

where,

$$\hat{x}(f) = \int_{-\infty}^{\infty} e^{-2\pi i f t} x(t) dt,$$

is the Fourier transform of the signal and f is the frequency in Hz, i.e., cycles per second. Often used is the angular frequency $\omega = 2\pi f$. Since the integral on the right-hand side is the energy of the signal, the integrand $| \hat{x}(f) |^2$ can be interpreted as a density function describing the energy per unit frequency contained in the signal at the frequency f. In light of this, the energy spectral density of a signal $x(t)$ is defined as:

$$S_{xx}(f) = \left| \hat{x}(f) \right|^2$$

As a physical example of how one might measure the energy spectral density of a signal, suppose $V(t)$ represents the potential (in volts) of an electrical pulse propagating along a transmission line of impedance Z, and suppose the line is terminated with a matched resistor (so that all of the pulse energy is delivered to the resistor and none is reflected back). By Ohm's law, the power delivered to the resistor at time t is equal to $V(t)^2 / Z$, so the total energy is found by integrating $V(t)^2 / Z$ with respect to time over the duration of the pulse. To find the value of the energy spectral density $S_{xx}(f)$ at frequency f, one could insert between the transmission line and the resistor a bandpass filter which passes only a narrow range of frequencies (Δf, say) near the frequency of interest and then measure the total energy $E(f)$ dissipated across the resistor. The value of the energy spectral density at f is then estimated to be $E(f) / \Delta f$. In this example, since the power $V(t)^2 / Z$ has units of $V^2\ \Omega^{-1}$, the energy $E(f)$ has units of $V^2\ s\ \Omega^{-1} = J$, and hence the estimate $E(f) / \Delta f$ of the energy spectral density has units of $J\ Hz^{-1}$, as required. In many situations, it is common to forgo the step of dividing by Z so that the energy spectral density instead has units of $V^2\ s\ Hz^{-1}$.

This definition generalizes in a straightforward manner to a discrete signal with an infinite number of values x_n such as a signal sampled at discrete times $x_n = x(n\Delta t)$:

$$S_{xx}(f) = (\Delta t)^2 \left| \sum_{n=-\infty}^{\infty} x_n e^{-2\pi i f n \Delta t} \right|^2 = \hat{x}_d(f) \hat{x}_d^*(f),$$

where $\hat{x}_d(f)$ is the discrete Fourier transform of x_n and $\hat{x}_d^*(f)$ is the complex conjugate of $\hat{x}_d(f)$ The sampling interval Δt is needed to keep the correct physical units and to ensure that we recover the continuous case in the limit $\Delta t \to 0$; however, in the mathematical sciences, the interval is often set to 1.

Power Spectral Density

The above definition of energy spectral density is suitable for transients (pulse-like signals) whose energy is concentrated around one time window; then the Fourier transforms of the signals generally exist. For continuous signals over all time, such as stationary processes, one must rather define the *power spectral density* (PSD); this describes how power of a signal or time series is distributed over frequency, Here, power can be the actual physical power, or more often, for convenience with abstract signals, is simply identified with the squared value of the signal. For example, statisticians study the variance of a function over time $x(t)$ (or over another independent variable), and using an analogy with electrical signals (among other physical processes), it is customary to refer to it as the *power spectrum* even when there is no physical power involved. If one were to create a physical voltage source which followed $x(t)$ and applied it to the terminals of a 1 ohm resistor, then indeed the instantaneous power dissipated in that resistor would be given by $x(t)^2$ watts.

The average power P of a signal $x(t)$ over all time is therefore given by the following time average:

$$P = \lim_{T\to\infty} \frac{1}{T}\int_0^T x(t)\,|^2\,dt\,.$$

Note that a stationary process, for instance, may have a finite power but an infinite energy. After all, energy is the integral of power, and the stationary signal continues over an infinite time. That is the reason that we cannot use the energy spectral density as defined above in such cases.

In analyzing the frequency content of the signal $x(t)$, one might like to compute the ordinary Fourier transform $\hat{x}(\omega)$; however, for many signals of interest the Fourier transform does not formally exist.[N 1] Because of this complication one can as well work with a truncated Fourier transform where the signal is integrated only over a finite interval $[0,T]$:

$$\hat{x}(\omega) = \frac{1}{\sqrt{T}}\int_0^T x(t)e^{-i\omega t}\,dt.$$

This is the amplitude spectral density. Then the power spectral density can be defined as:

$$S_{xx}(\omega) = \lim_{T\to\infty} \mathrm{E}\left[\left|\hat{x}(\omega)\right|^2\right]$$

Here E denotes the expected value; explicitly, we have,

$$[|\,\hat{x}(\omega)\,|^2] = \mathrm{E}[\frac{1}{T}\int_0^T x^*(t)e^{i\omega t}\,dt\int_0^T x(t')e^{-i\omega t'}\,dt'] = \frac{1}{T}\int_0^T\int_0^T \mathrm{E}[x^*(t)x(t')]e^{i\omega(t-t')}\,dt\,dt'.$$

In the latter form (for a stationary random process), one can make the change of variables

$\Delta t = t - t'$ and with the limits of integration (rather than $[0, T]$) approaching infinity, the resulting power spectral density $S_{xx}(\omega)$ and the autocorrelation function of this signal are seen to be Fourier transform pairs (Wiener–Khinchin theorem). The autocorrelation function is a statistic defined as

$$R_{xx}(\tau) = \langle X(t)X(t+\tau)\rangle = \mathrm{E}[X(t)X(t+\tau)]$$

or more generally as

$$R_{xx}(\tau) = \langle X(t)X(t-\tau)^*\rangle = \langle X(t)^* X(t+\tau)\rangle$$

in the case that $X(t)$ is complex-valued. Provided that $R_{xx}(\tau)$ is absolutely integrable (which is not always true),

$$S_{xx}(\omega) = \int_{-\infty}^{\infty} R_{xx}(\tau)e^{-i\omega\tau}\,d\tau = \hat{R}_{xx}(\omega)$$

Many authors use this equality to actually *define* the power spectral density.

The power of the signal in a given frequency band $[f_1, f_2]$ (or $[\omega_1, \omega_2]$) can be calculated by integrating over frequency. Since $S_{xx}(-\omega) = S_{xx}(\omega)$, an equal amount of power can be attributed to positive and negative frequencies, which accounts for the factor of 2 in the following form (such trivial factors dependent on conventions used):

$$P_{\text{bandlimited}} = 2\int_{f_1}^{f_2} S_{xx}(2\pi f)\,df = \frac{1}{\pi}\int_{\omega_1}^{\omega_2} S_{xx}(\omega)\,d\omega$$

More generally, similar techniques may be used to estimate a time-varying spectral density. In this case the truncated Fourier transform defined above over the finite time interval $(0, T)$ is *not* evaluated in the limit of T approaching infinity. This results in decreased spectral coverage and resolution since frequencies of less than $1/T$ are not sampled, and results at frequencies which are not an integer multiple of $1/T$ are not independent. Just using a single such time series, the estimated power spectrum will be very "noisy"; however this can be alleviated if it is possible to evaluate the expected value (in the above equation) using a large (or infinite) number of short-term spectra corresponding to statistical ensembles of realizations of $x(t)$ evaluated over the specified time window.

This definition of the power spectral density can be generalized to discrete time variables x_n. As above we can consider a *finite* window of $1 \le n \le N$ with the signal sampled at discrete times $x_n = x(n\Delta t)$ for a total measurement period $T = N\text{Ä}t$. Then a single estimate of the PSD can be obtained through summation rather than integration:

$$\tilde{S}_{xx}(\omega) = \frac{(\Delta t)^2}{T}\left|\sum_{n=1}^{N} x_n e^{-i\omega n\text{Ä}t}\right|^2 .$$

As before, the actual PSD is achieved when N (and thus T) approach infinity and the expected value is formally applied. In a real-world application, one would typically average this single-measurement PSD over many trials to obtain a more accurate estimate of the theoretical PSD of the physical process underlying the individual measurements. This computed PSD is sometimes called a periodogram. This periodogram converges to the true PSD as the number of estimates as well as the averaging time interval T approach infinity (Brown & Hwang).

If two signals both possess power spectral densities, then the #Cross-spectral density can similarly be calculated; as the PSD is related to the autocorrelation, so is the cross-spectral density related to the cross-correlation.

Properties of the Power Spectral Density

Some properties of the PSD include:

- The spectrum of a real valued process (or even a complex process using the above definition) is real and an even function of frequency: $S_{xx}(-\omega) = S_{xx}(\omega)$.
- If the process is continuous and purely indeterministic, the autocovariance function can be reconstructed by using the Inverse Fourier transform.
- The PSD can be used to compute the variance (net power) of a process by integrating over frequency:

$$\text{Var}(X_n) = \frac{1}{\pi}\int_0^{\infty} S_{xx}(\omega)d\omega.$$

- Being based on the Fourier transform, the PSD is a linear function of the autocovariance function in the sense that if R_{xx} is decomposed into two functions.

$$R_{xx}(\tau) = \alpha_1 R_{xx,1}(\tau) + \alpha_2 R_{xx,2}(\tau),$$

then,

$$S_{xx}(\omega) = \alpha_1 S_{xx,1}(\omega) + \alpha_2 S_{xx,2}(\omega)$$

The *integrated spectrum* or *power spectral distribution* $F(\omega)$ is defined as,

$$F(\omega) = \int_{-\infty}^{\omega} S_{xx}(\omega')d\omega'.$$

Cross Power Spectral Density

Given two signals $x(t)$ and $y(t)$, each of which possess power spectral densities $S_{xx}(\omega)$ and $S_{yy}(\omega)$,it is possible to define a cross power spectral density (CPSD) or cross spectral density (CSD) given by:

$$S_{xy}(\omega) = \lim_{T\to\infty} \mathrm{E}\left[\left(F_x^T(\omega)\right)^* F_y^T(\omega)\right].$$

The cross-spectral density (or 'cross power spectrum') is thus the Fourier transform of the cross-correlation function.

$$S_{xy}(\omega) = \int_{-\infty}^{\infty} R_{xy}(t) e^{-j\omega t} dt = \int_{-\infty}^{\infty} [\int_{-\infty}^{\infty} x(\tau) \cdot y(\tau + t) d\tau] e^{-j\omega t} dt,$$

where $R_{xy}(t)$ is the cross-correlation of $x(t)$ and $y(t)$.

By an extension of the Wiener–Khinchin theorem, the Fourier transform of the cross-spectral density $S_{xy}(\omega)$ is the cross-covariance function. In light of this, the PSD is seen to be a special case of the CSD for $x(t) = y(t)$.

For discrete signals x_n and y_n, the relationship between the cross-spectral density and the cross-co-variance is,

$$S_{xy}(\omega) = \frac{1}{2\pi} \sum_{n=-\infty}^{\infty} R_{xy}(n) e^{-j\omega n}$$

Estimation

The goal of spectral density estimation is to estimate the spectral density of a random signal from a sequence of time samples. Depending on what is known about the signal, estimation techniques can involve parametric or non-parametric approaches, and may be based on time-domain or frequency-domain analysis. For example, a common parametric technique involves fitting the observations to an autoregressive model. A common non-parametric technique is the periodogram.

The spectral density is usually estimated using Fourier transform methods (such as the Welch method), but other techniques such as the maximum entropy method can also be used.

Properties

- The spectral density of $f(t)$ and the autocorrelation of $f(t)$ form a Fourier transform pair (for PSD versus ESD, different definitions of autocorrelation function are used). This result is known as Wiener–Khinchin theorem.
- One of the results of Fourier analysis is Parseval's theorem which states that the area under the energy spectral density curve is equal to the area under the square of the magnitude of the signal, the total energy:

$$\int_{-\infty}^{\infty} |f(t)|^2 dt = \int_{-\infty}^{\infty} ESD(\omega) d\omega.$$

The above theorem holds true in the discrete cases as well. A similar result holds for power: the area under the power spectral density curve is equal to the total signal power, which is $R(0)$, the autocorrelation function at zero lag. This is also (up to a constant which depends on the normalization factors chosen in the definitions employed) the variance of the data comprising the signal.

Related Concepts

- The spectral centroid of a signal is the midpoint of its spectral density function, i.e. the frequency that divides the distribution into two equal parts.
- The spectral edge frequency of a signal is an extension to any proportion instead of two equal parts.
- The spectral density is a function of frequency, not a function of time. However, the spectral density of small windows of a longer signal may be calculated, and plotted versus time associated with the window. Such a graph is called a *spectrogram*. This is the basis of a number of spectral analysis techniques such as the short-time Fourier transform and wavelets.
- A "spectrum" generally means the power spectral density, which depicts the distribution of signal content over frequency. This is not to be confused with the frequency response of a transfer function which also includes a phase (or equivalently, a real and imaginary part as a function of frequency). For transfer functions, (e.g., Bode plot, chirp) the complete frequency response may be graphed in two parts, amplitude versus frequency and phase versus frequency (or less commonly, as real and imaginary parts of the transfer function). The impulse response (in the time domain) $h(t)$, cannot generally be uniquely recovered from the amplitude spectral density part alone without the phase function. Although these are also Fourier transform pairs, there is no symmetry (as there is for the autocorrelation) forcing the Fourier transform to be real-valued.

Applications

Electrical Engineering

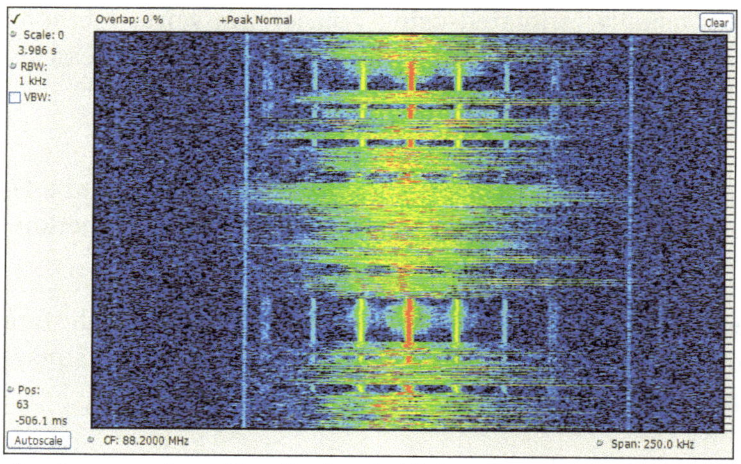

Spectrogram of an FM radio signal with frequency on the horizontal axis and time increasing upwards on the vertical axis.

The concept and use of the power spectrum of a signal is fundamental in electrical engineering, especially in electronic communication systems, including radio communications, radars, and related systems, plus passive remote sensing technology. Electronic instruments called spectrum analyzers are used to observe and measure the *power spectra* of signals.

The spectrum analyzer measures the magnitude of the short-time Fourier transform (STFT) of an input signal. If the signal being analyzed can be considered a stationary process, the STFT is a good smoothed estimate of its power spectral density.

Spectral Density Estimation

In statistical signal processing, the goal of spectral density estimation (SDE) is to estimate the spectral density (also known as the power spectral density) of a random signal from a sequence of time samples of the signal. Intuitively speaking, the spectral density characterizes the frequency content of the signal. One purpose of estimating the spectral density is to detect any periodicities in the data, by observing peaks at the frequencies corresponding to these periodicities.

Some SDE techniques assume that a signal is composed of a limited (usually small) number of generating frequencies plus noise and seek to find the location and intensity of the generated frequencies. Others make no assumption on the number of components and seek to estimate the whole generating spectrum.

Techniques

Many other techniques for spectral estimation have been developed to mitigate the disadvantages of the basic periodogram. These techniques can generally be divided into *non-parametric* and *parametric* methods. The non-parametric approaches explicitly estimate the covariance or the spectrum of the process without assuming that the process has any particular structure. Some of the most common estimators in use for basic applications (e.g. Welch's method) are non-parametric estimators closely related to the periodogram. By contrast, the parametric approaches assume that the underlying stationary stochastic process has a certain structure that can be described using a small number of parameters (for example, using an auto-regressive or moving average model). In these approaches, the task is to estimate the parameters of the model that describes the stochastic process.

Following is a partial list of non-parametric spectral density estimation techniques:

- Periodogram, the modulus-squared of the discrete Fourier transform.
- Bartlett's method is the average of the periodograms taken of multiple segments of the signal to reduce variance of the spectral density estimate.
- Welch's method a windowed version of Bartlett's method that uses overlapping segments.
- Multitaper is a periodogram-based method that uses multiple tapers, or windows, to form independent estimates of the spectral density to reduce variance of the spectral density estimate.
- Least-squares spectral analysis, based on least squares fitting to known frequencies.
- Non-uniform discrete Fourier transform is used when the signal samples are unevenly spaced in time.

- Singular spectrum analysis is a nonparametric method that uses a singular value decomposition of the covariance matrix to estimate the spectral density.
- Short-time Fourier transform.
- Critical filter is a nonparametric method based on information field theory that can deal with noise, incomplete data, and instrumental response functions.

Below is a partial list of parametric techniques:

- Autoregressive model (AR) estimation, which assumes that the nth sample is correlated with the previous p samples.
- Moving-average model (MA) estimation, which assumes that the nth sample is correlated with noise terms in the previous p samples.
- Autoregressive moving average (ARMA) estimation, which generalizes the AR and MA models.
- MUltiple SIgnal Classification (MUSIC) is a popular superresolution method.
- Maximum entropy spectral estimation is an *all-poles* method useful for SDE when singular spectral features, such as sharp peaks, are expected.

Parametric Estimation

In parametric spectral estimation, one assumes that the signal is modeled by a stationary process which has a spectral density function (SDF) $S(f;a_1,\ldots,a_p)$ that is a function of the frequency f and p parameters $a_1,\ldots,a_p$. The estimation problem then becomes one of estimating these parameters.

The most common form of parametric SDF estimate uses as a model an autoregressive model $\mathrm{AR}(p)$ of order p. A signal sequence $\{Y_t\}$ obeying a zero mean $\mathrm{AR}(p)$ process satisfies the equation:

$$Y_t = \phi_1 Y_{t-1} + \phi_2 Y_{t-2} + \cdots + \phi_p Y_{t-p} + \epsilon_t,$$

where the $\phi_1,\ldots,\phi_p$ are fixed coefficients and ϵ_t is a white noise process with zero mean and *innovation variance* σ_p^2. The SDF for this process is,

$$S(f;\phi_1,\ldots,\phi_p,\sigma_p^2) = \frac{\sigma_p^2 \Delta t}{|1-\sum_{k=1}^{p}\phi_k e^{-2i\pi f k \Delta t}|^2} \quad |f| \le f_N,$$

with Δt the sampling time interval and f_N the Nyquist frequency.

There are a number of approaches to estimating the parameters $\phi_1,\ldots,\phi_p,\sigma_p^2$ of the $\mathrm{AR}(p)$ process and thus the spectral density:

- The *Yule-Walker estimators* are found by recursively solving the Yule-Walker equations for an $\mathrm{AR}(p)$ process.

- The *Burg estimators* are found by treating the Yule-Walker equations as a form of ordinary least squares problem. The Burg estimators are generally considered superior to the Yule-Walker estimators. Burg associated these with maximum entropy spectral estimation.
- The *forward-backward least-squares estimators* treat the $\mathrm{AR}(p)$ process as a regression problem and solves that problem using forward-backward method. They are competitive with the Burg estimators.
- The *maximum likelihood estimators* estimate the parameters using a maximum likelihood approach. This involves a nonlinear optimization and is more complex than the first three.

Alternative parametric methods include fitting to a moving average model (MA) and to a full autoregressive moving average model (ARMA).

Frequency Estimation

Frequency estimation is the process of estimating the complex frequency components of a signal in the presence of noise given assumptions about the number of the components. This contrasts with the general methods above, which do not make prior assumptions about the components.

Finite Number of Tones

A typical model for a signal $x(n)$ consists of a sum of p complex exponentials in the presence of white noise, $w(n)$

$$x(n) = \sum_{i=1}^{p} A_i e^{jn\omega_i} + w(n).$$

The power spectral density of $x(n)$ is composed of p impulse functions in addition to the spectral density function due to noise.

The most common methods for frequency estimation involve identifying the noise subspace to extract these components. These methods are based on eigen decomposition of the autocorrelation matrix into a signal subspace and a noise subspace. After these subspaces are identified, a frequency estimation function is used to find the component frequencies from the noise subspace. The most popular methods of noise subspace based frequency estimation are Pisarenko's method, the multiple signal classification (MUSIC) method, the eigenvector method, and the minimum norm method.

Pisarenko's method,

$$\hat{P}_{\mathrm{PHD}}\left(e^{j\omega}\right) = \frac{1}{\left|\mathrm{e}^{H}\mathrm{v}_{\min}\right|^{2}}$$

Music,

$$\hat{P}_{\mathrm{MU}}(e^{j\omega}) = \frac{1}{\sum_{i=p+1}^{M} |\mathrm{e}^{H}\mathrm{v}_i|^2}$$

Eigenvector method,

$$\hat{P}_{\text{EV}}\left(e^{j\omega}\right)=\frac{1}{\sum_{i=p+1}^{M}\frac{1}{\lambda_i}\left|\mathrm{e}^H\mathrm{v}_i\right|^2}$$

Minimum norm method,

$$\hat{P}_{\text{MN}}\left(e^{j\omega}\right)=\frac{1}{\left|\mathrm{e}^H\mathrm{a}\right|^2};\ \mathrm{a}=\lambda\mathrm{P}_n\mathrm{u}_1$$

Single Tone

If one only wants to estimate the single loudest frequency, one can use a pitch detection algorithm. If the dominant frequency changes over time, then the problem becomes the estimation of the instantaneous frequency as defined in the time–frequency representation. Methods for instantaneous frequency estimation include those based on the Wigner-Ville distribution and higher order ambiguity functions.

If one wants to know *all* the (possibly complex) frequency components of a received signal (including transmitted signal and noise), one uses a discrete Fourier transform or some other Fourier-related transform.

Example Calculation

Suppose x_n, from $n=0$ to $N-1$ is a time series (discrete time) with zero mean. Suppose that it is a sum of a finite number of periodic components (all frequencies are positive):

$$\begin{aligned} x_n &= \sum_k A_k \cdot \sin(2\pi\nu_k n+\phi_k) \\ &= \sum_k\left(\overbrace{a_k}^{A_k\sin(\phi_k)}\cos(2\pi\nu_k n)+\overbrace{b_k}^{A_k\cos(\phi_k)}\sin(2\pi\nu_k n)\right)\end{aligned}$$

The variance of x_n is, for a zero-mean function as above, given by $\frac{1}{N}\sum_{n=0}^{N-1}x_n^2$. If these data were samples taken from an electrical signal, this would be its average power (power is energy per unit time, so it is analogous to variance if energy is analogous to the amplitude squared).

Now, for simplicity, suppose the signal extends infinitely in time, so we pass to the limit as $N\to\infty$. If the average power is bounded, which is almost always the case in reality, then the following limit exists and is the variance of the data.

$$\lim_{N\to\infty}\frac{1}{N}\sum_{n=0}^{N-1}x_n^2.$$

Again, for simplicity, we will pass to continuous time, and assume that the signal extends infinitely in time in both directions. Then these two formulas become,

$$x(t) = \sum_k A_k \cdot \sin(2\pi \nu_k t + \phi_k)$$

and

$$\lim_{T \to \infty} \frac{1}{2T} \int_{-T}^{T} x(t)^2 \, dt.$$

The root mean square of $1/\sqrt{2}$ is, so the variance of $A_k \sin(2\pi\nu_k t + \phi_k)$ is $A_k^2 / 2$. Hence, the contribution to the average power of $x(t)$ coming from the component with frequency ν_k is $A_k^2 / 2$. All these contributions add up to the average power of.

Then the power as a function of frequency is $A_k^2 / 2$, and its statistical cumulative distribution function $S(\nu)$ will be,

$$S(\nu) = \sum_{k:\nu_k < \nu} \frac{1}{2} A_k^2.$$

S is a step function, monotonically non-decreasing. Its jumps occur at the frequencies of the periodic components of x, and the value of each jump is the power or variance of that component.

The variance is the covariance of the data with itself. If we now consider the same data but with a lag of τ, we can take the covariance of $x(t)$ with $x(t+\tau)$, and define this to be the autocorrelation function c of the signal (or data) x:

$$c(\tau) = \lim_{T \to \infty} \frac{1}{2T} \int_{-T}^{T} (t)x(t+\tau) dt.$$

If it exists, it is an even function of τ. If the average power is bounded, then c exists everywhere, is finite, and is bounded by $c(0)$, which is the average power or variance of the data.

It can be shown that c can be decomposed into periodic components with the same periods as x:

$$c(\tau) = \sum_k \frac{1}{2} A_k^2 \cos(2\pi \nu_k \tau).$$

This is in fact the spectral decomposition of c over the different frequencies, and is related to the distribution of power of x over the frequencies: the amplitude of a frequency component of c is its contribution to the average power of the signal.

The power spectrum of this example is not continuous, and therefore does not have a derivative, and therefore this signal does not have a power spectral density function. In general, the power spectrum will usually be the sum of two parts: a line spectrum such as in this example, which is not continuous and does not have a density function, and a residue, which is absolutely continuous and does have a density function.

Frequency Spectrum Analysis

Whether it is for tuning of a musical instrument or searching for extraterrestrial radio signals, frequency spectrum analysis of signals provides valuable information otherwise not available. While an oscilloscope displays a signal with respect to time, a spectrum analyzer plots it with respect to frequency. In theory, Fourier theorem states that a signal is composed of a number of sinusoidal signals.

Analyzing the amplitude, frequency, and phase of these sinusoidal signals is referred to as the frequency spectrum analysis of the signal. To extract these parameters, the signal is filtered, digitized, and Fourier transformed to the frequency domain. The amplitude and the phase spectrum of the signal can then be plotted based on these parameters.

A spectrum analyzer performs the described operation on its input signal and provides the spectrum on its display. Figure shows the amplitude spectrum of a 10 MHz 1 V or .707 VRMS sine wave on Analog Arts SL987 spectrum analyzer.

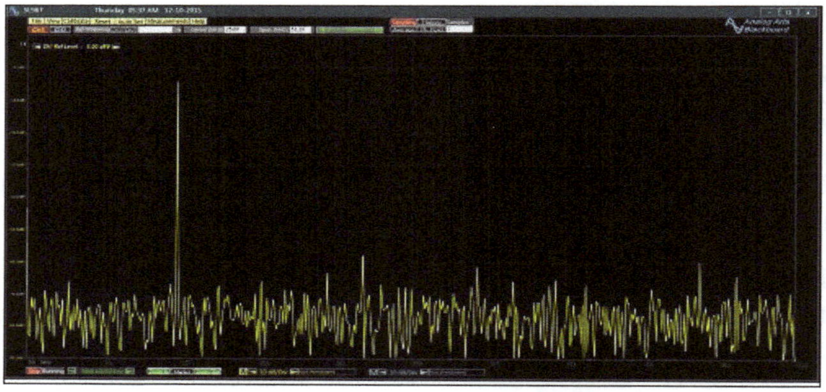

The amplitude spectrum of a 10MHz sine wave.

In real life, signals are more complex than a single tone sine wave. Consider an antenna signal coming from an AM radio station. Looking at this signal on an oscilloscope does not offer any meaningful information. A spectrum analyzer, however, provides the composition of the signal. It clearly shows the carrier signal as well as the modulating signal.

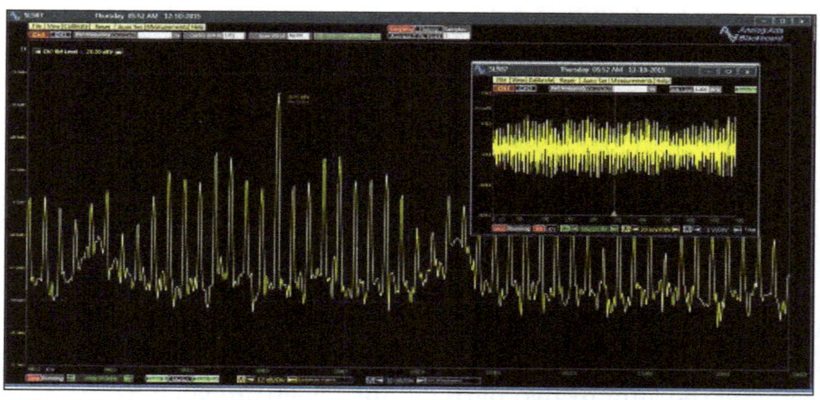

The time domain and frequency spectrum representation of an AM signal.

In figure illustrates the time domain oscilloscope plot and the corresponding frequency amplitude spectrum. To an initiated engineer, the AM nature of the signal is evident from the spectrum plot. Based on this fact, further analysis could follow. In Fact, a frequency spectrum plot can reveal the nature of each individual signal present on the antenna whether they are coming from a radio station or an extraterrestrial source.

Spectrum analysis of signals is particularly helpful in the design and developments of an electronic product. A frequency plot can immediately expose signal distortion through various stages of a design. Figure compares the spectrum of a sine wave signal before and after passing through a signal processing stage of a design. The processed signal, shown in blue, is clearly distorted by the stage. Probing different sections of the design by the spectrum analyzer can pin point the cause of the distortion whether it is a component issue or grounding and power supply problems.

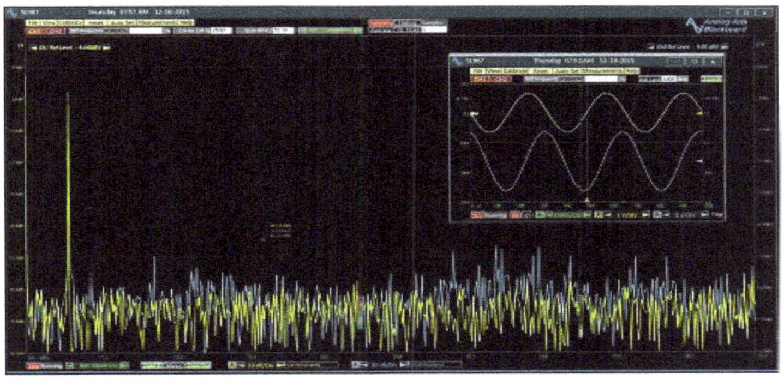

The spectrum of a sine wave before and after going through a design stage.

Another example of frequency signal analysis is the evaluation of filters. A filter by definition limits the frequency performance of a system to a certain bandwidth. The characteristic of the filter in its pass-band, stop-band, and the transition between these two areas defines the quality of the filter. An easy way to evaluate the characteristic of the filter is applying a white noise signal to the input of the filter and looking at its output frequency spectrum.

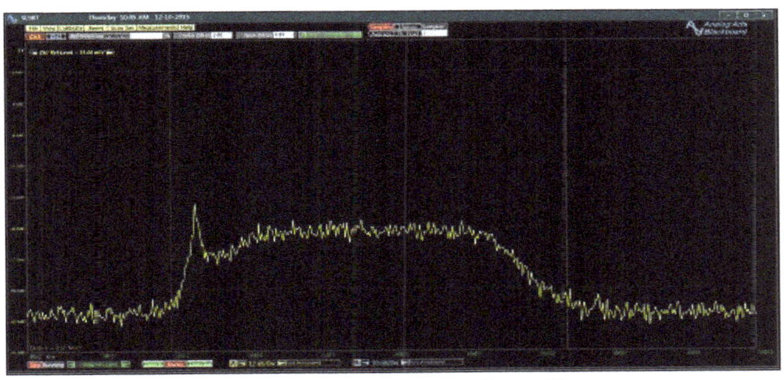

The spectrum plot of a bandpass filter.

There are numerous other applications for frequency spectrum analysis. Traditionally, the technique has been widely applied in fields such as radar, communications, and astronomy. As the prices of spectrum analyzers are coming down, it is finding new applications in other areas. This popularity is only going to grow over time for its unique ability to reveal the composition of signals and provide a library of various components that make up the signal.

References

- Dongarra, Jack; Sullivan, Francis (January 2000). "Guest Editors Introduction to the top 10 algorithms". Computing in Science Engineering. 2 (1): 22–23. Doi:10.1109/MCISE.2000.814652. ISSN 1521-9615
- Kent, Ray D.; Read, Charles (2002). Acoustic Analysis of Speech. ISBN 0-7693-0112-6. ISBN 978-0-7693-0112-9
- Rockmore, Daniel N. (January 2000). "The FFT: an algorithm the whole family can use". Computing in Science Engineering. 2 (1): 60–64. Citeseerx 10.1.1.17.228. Doi:10.1109/5992.814659. ISSN 1521-9615
- Storch, H. Von; F. W Zwiers (2001). Statistical analysis in climate research. Cambridge University Press. ISBN 978-0-521-01230-0
- Frequency-spectrum-analysis-signals, technology-trends: electronicsforu.com, Retrieved 17 May, 2019

5

Z-Transform

The Z-Transform is used to convert a discrete-time signal into a complex frequency-domain representation. Some of its important properties are linearity, time shifting, time expansion, convolution and time accumulation. This chapter closely examines these properties and related aspects of Z-Transform to provide an extensive understanding of the subject.

Z-Transform of a signal provides a valuable technique for analysis and design of the discrete time signal and discrete-time LTI system.

Z-Transform of a discrete time signal has both imaginary and real part. The plot of the imaginary part versus real part is called as the z plane. The poles and zeros of the discrete time signals are plotted in the complex z plane. Pole-zero plot is the main characteristics of the discrete time signals. Using pole-zero plot we can check the stability of the system which we will see in the upcoming posts.

Advantages of the Z-Transform

Following are some of the main advantages of the Z-Transform:

- We can simplify the solution of a differential equation using Z-Transform.
- By the use of Z-Transform, we can completely characterize given discrete time signals and LTI systems.
- The stability of the LTI system can be determined using a Z-Transform.
- Mathematical calculations can be reduced by using the Z-Transform.

  For example, the convolution operation is transformed into a simple multiplication operation.

Z-Transforms

There are two types of Z-Transform:

- Single sided Z-Transform.
- Double sided Z-Transform.

Single Sided Z-Transform

Single sided Z-Transform can be defined as,

$$X(Z) = \sum_{n=0}^{\infty} x(n) Z^{-n}$$

Double Sided Z-Transform

Double sided Z-Transform can be defined as,

$$X(Z) = \sum_{n=-\infty}^{\infty} x(n) Z^{-n}$$

In single sided Z-Transform only positive values of n are used (from 0 to 8) hence called single sided, whereas in the double sided values of n are ranging from -8 to +8.

Representation of the Z-Transforms

Z-Transform of the signal x(n) is represented as,

$$\mathrm{X(Z)} = Z\{x(n)\}$$

The relation between x(n) and X(Z) is denoted as follows:

$$\mathrm{x(n)} = \overset{Z}{\longleftrightarrow} \mathrm{X(Z)}$$

Where X(Z) is the Z-Transform of the signal x(n). The arrow is bidirectional which indicates that we can obtain x(n) from X(Z) also, which is called as inverse Z-Transform.

x(n) and X(Z) is called as Z-Transform pair.

Properties of Z-Transform

The z-transform has a set of properties in parallel with that of the Fourier transform (and Laplace transform). The difference is that we need to pay special attention to the ROCs. In the following, we always assume,

$$\mathcal{Z}[x[n]] = X(z) \quad ROC = R_x$$

and

$$\mathcal{Z}[y[n]] = Y(z) \quad ROC = R_y$$

Linearity

$$\mathcal{Z}[ax[n] + by[n]] = aX(z) + bY(z), \quad ROC \supseteq (R_x \cap R_y)$$

While it is obvious that the ROC of the linear combination of $x[n]$ and $y[n]$ should be the intersection of the their individual ROCs $R_x \cap R_y$. in which both $X(z)$ and $Y(z)$ exist, note that in some cases the ROC of the linear combination could be larger than $R_x \cap R_y$. For example, for both $x[n] = a^n u[n]$ and $y[n] = a^n u[n-1]$, the ROC is $|z| \quad |a|$, but the ROC of their difference $a^n u[n] - a^n u[n-1] = \delta[n]$ is the entire z-plane.

Time Shifting

$$\mathcal{Z}[x[n-n_0]] = z^{-n_0} X(z), \quad ROC = R_x$$

Proof:

$$\mathcal{Z}[x[n-n_0]] = \sum_{n=-\infty}^{\infty} x[n-n_0] z^{-n}$$

Define $m = n - n_0$, we have $n = m + n_0$ and

$$\sum_{m=-\infty}^{\infty} x[m] z^{-m} z^{-n_0} = z^{-n_0} X(z)$$

The new ROC is the same as the old one except the possible addition/deletion of the origin or infinity as the shift may change the duration of the signal.

Time Expansion (Scaling)

$$\mathcal{Z}[x[n/k]] = X(z^k), \quad ROC = R_x^{1/k}$$

The discrete signal $x[n]$ cannot be continuously scaled in time as n has to be an integer (for a non-integer $n x[n]$ is zero). Therefore $x[n/k]$ is defined as:

$$x[n/k] \stackrel{\Delta}{=} \begin{cases} x[n/k] & \text{if } n \text{ is a multiple of } k \\ 0 & \text{else} \end{cases}$$

Example: If $x[n]$ is ramp,

n	1	2	3	4	5	6
$x[n]$	1	2	3	4	5	6

then the expanded version $x[n/2]$ is,

n	1	2	3	4	5	6
$n/2$	0.5	1	1.5	2	2.5	3
m		1		2		3
$x[n/2]$	0	1	0	2	0	3

where m is the integer part of n/k.

Proof: The z-transform of such an expanded signal is,

$$\mathcal{Z}[x[n/k]] = \sum_{n=-\infty}^{\infty} x[n/k]z^{-n} = \sum_{m=-\infty}^{\infty} x[m]z^{-km} = X(z^k)$$

Note that the change of the summation index from n to m has no effect as the terms skipped are all zeros.

Convolution

$$\mathcal{Z}[x[n]*y[n]] = X(z)Y(z), \quad ROC \supseteq (R_x \cap R_y)$$

The ROC of the convolution could be larger than the intersection of R_x and R_y, due to the possible pole-zero cancellation caused by the convolution.

Time Difference

$$\mathcal{Z}[x[n]-x[n-1]] = (1-z^{-1})X(z), \quad ROC = R_x$$

Proof:

$$\mathcal{Z}[x[n]-x[n-1]] = X(z) - z^{-1}X(z) = (1-z^{-1})X(z) = \frac{z-1}{z}X(z)$$

Note that due to the additional zero $z=1$ and pole $z=0$, the resulting ROC is the same as R_x except the possible deletion of $z=0$ caused by the added pole and addition of $z=1$ caused by the added zero which may cancel an existing pole.

Time Accumulation

$$\mathcal{Z}[\sum_{k=-\infty}^{n} x[k]] = \frac{1}{1-z^{-1}}X(z), \quad ROC \supseteq [R_x \cap (|z|>1)]$$

Proof: The accumulation of $x[n]$ can be written as its convolution with $u[n]$:

$$u[n]*x[n] = \sum_{k=-\infty}^{\infty} u[n-k]x[k] = \sum_{k=-\infty}^{n} x[k]$$

Applying the convolution property, we get

$$\mathcal{Z}[\sum_{k=-\infty}^{n} x[k]] = Z[u[n]*x[n]] = \frac{1}{1-z^{-1}}X(z)$$

as $\mathcal{Z}[u[n]] = 1/(1-z^{-1})$.

Time Reversal

$$\mathcal{Z}[x[-n]] = X(1/z) \quad ROC = 1/R_x$$

Proof:

$$\mathcal{Z}[x[-n]] = \sum_{n=-\infty}^{\infty} x[-n]z^{-n} = \sum_{m=-\infty}^{\infty} x[m](\frac{1}{z})^{-m} = X(1/z)$$

where $m = -n$.

Scaling in Z-domain

$$\mathcal{Z}[a^n x[n]] = X\left(\frac{z}{a}\right), \quad ROC = |a| R_x$$

Proof:

$$\mathcal{Z}[a^n x[n]] = \sum_{n=-\infty}^{\infty} x[n]\left(\frac{z}{a}\right)^{-n} = X\left(\frac{z}{a}\right)$$

In particular, if $a = e^{j\omega_0}$,the above becomes,

$$\mathcal{Z}[e^{jn\omega_0} x[n]] = X(e^{-j\omega_0} z) \quad ROC = R_x$$

The multiplication by $e^{-j\omega_0}$ to $\underline{z}$ corresponds to a rotation by angle ω_0 in the z-plane, i.e., a frequency shift by ω_0. The rotation is either clockwise $(\omega_0 > 0)$ or counter clockwise $(\omega_0 < 0)$ corresponding to, respectively, either a left-shift or a right shift in frequency domain. The property is essentially the same as the frequency shifting property of discrete Fourier transform.

Conjugation

$$\mathcal{Z}[x^*[n]] = X^*(z^*), \quad ROC = R_x$$

Proof: Complex conjugate of the z-transform of $x[n]$ is,

$$X^*(z) = [\sum_{n=-\infty}^{\infty} x[n]z^{-n}]^* = \sum_{n=-\infty}^{\infty} x^*[n](z^*)^{-n}$$

Replacing z by z^*, we get the desired result.

Differentiation in Z-domain

$$\mathcal{Z}[n\,x[n]] = -z\frac{d}{dz}X(z), \quad ROC = R_x$$

Proof:

$$\frac{d}{dz}X(z)=\sum_{n=-\infty}^{\infty}x[n]\frac{d}{dz}(z^{-n})=\sum_{n=-\infty}^{\infty}(-n)x[n]z^{-n-1}=\frac{-1}{z}\sum_{n=-\infty}^{\infty}nx[n]z^{-n}$$

i.e.,

$$\mathcal{Z}[nx[n]]=-z\frac{d}{dz}X(z)$$

Example:

Taking derivative with respect to z of the right side of,

$$\mathcal{Z}[a^n u[n]]=\frac{1}{1-az^{-1}} \quad |z|>|a|$$

we get,

$$\frac{d}{dz}\left[\frac{1}{1-az^{-1}}\right]=\frac{-az^{-2}}{(1-az^{-1})^2}$$

Due to the property of differentiation in z-domain, we have:

$$\mathcal{Z}[na^n u[n]]=\frac{az^{-1}}{(1-az^{-1})^2} \quad |z|>|a|$$

Note that for a different ROC $|z|<|a|$, we have:

$$\mathcal{Z}[-na^n u[-n-1]]=\frac{a\,z^{-1}}{(1-a\,z^{-1})^2} \quad |z|<|a|$$

Inverse Z-Transform

The inverse Z-Transforms can be obtained by using any one of the following methods. They are

- Power series method.
- Partial fraction method.
- Inversion Integral method.
- Long division method.

Power Series Method

This is the simplest method of finding inverse $Z-$transform. Here F(z) can the be expanded in a series of ascending powers and the coefficient of z^{-n} will be the of z desired inverse $Z-$ transform.

Example:

Find the inverse $Z-$transform of $\log\{z\ /(z+1)\}$ by power series method.

$$\text{Putting } z=\frac{1}{y}, F(z)=\log\left\{\frac{1/y}{(1/y)+1}\right\}$$

$$=\log\left\{\frac{1}{1+y}\right\}$$

$$=-\log(1+y)$$

$$=-y+\frac{y^2}{2}-\frac{y^3}{3}+\ldots\ldots.$$

$$=-z^{-1}+\frac{1}{2}z^{-2}-\frac{1}{3}z^{-3}+\ldots..+\frac{(-1)^n}{n}z^{-n}$$

$$\text{Thus, } f_n=\begin{cases}0,\ \text{for}\ n=0\\ (-1)^n/n,\ \text{otherwise}\end{cases}$$

Partial Fraction Method

Here, $F(z)$ is resolved into partial fractions and the inverse transform can be taken directly.

Example:

Find the inverse $Z-$ transform of,

$$\frac{z}{z^2+7z+10}$$

$$\text{Let } F(z)=\frac{z}{z^2+7z+10}$$

$$\text{Then } \frac{F(z)}{z}=\frac{1}{z^2+7z+10}=\frac{1}{(z+2)(z+5)}$$

$$\text{Now, consider } \frac{1}{(z+2)(z+5)}=\frac{A}{z+2}+\frac{B}{z+5}$$

$$=\frac{1}{3}\frac{1}{z+2}-\frac{1}{3}\frac{1}{z+5}$$

$$\text{Therefore, } F(z)=\frac{1}{3}\frac{z}{z+2}-\frac{1}{3}\frac{z}{z+5}$$

Inverting, we get

$$=\frac{1}{3}(-2)^n-\frac{1}{3}(-5)^n$$

Example:

Find the inverse Z – trnsform of $\dfrac{8z^2}{(2z-1)\,(4z-1)}$

$$\text{Let } F(z) = \frac{8z^2}{(2z-1)(4z-1)} = \frac{z^2}{(z-1\frac{1}{2})(z-\frac{1}{4})}$$

$$\text{Then } \frac{F(z)}{z} = \frac{1}{(z-\frac{1}{2})(z-\frac{1}{4})}$$

Now, $\dfrac{z}{(z+\frac{1}{2})(z+\frac{1}{4})} = \dfrac{A}{z-\frac{1}{2}} + \dfrac{B}{z-\frac{1}{4}}$

We get, $\dfrac{F(z)}{z} = \dfrac{2}{z-\frac{1}{2}} - \dfrac{1}{z-\frac{1}{4}}$

Therefore, $F(z) = 2\dfrac{z}{z-\frac{1}{2}} - \dfrac{z}{z-\frac{1}{4}}$

Inverting, we get

$$f_n = Z^{-1}\{F(z)\} = 2Z^{-1}\left\{\frac{z}{z-\frac{1}{2}}\right\} - Z^{-1}\left\{\frac{z}{z-\frac{1}{4}}\right\}$$

i.e, $f_n = 2(1/2)^n - (1/4)^n, \; n = 0,1,2,......$

Example:

Find, $Z^{-1}\left\{\dfrac{4-8z^{-1}+6z^{-2}}{(1+z^{-1})(1-2z^{-1})^2}\right\}$ by the method of partial fractions.

Let, $F(z) = \dfrac{4-8z^{-1}+6z^{-2}}{(1+z^{-1})(1-2z^{-1})^2}$

$$= \frac{4z^3-8z^2+6z}{(z+1)(z-2)^2}$$

Then, $\dfrac{F(z)}{z} = \dfrac{4z^3-8z+6z}{(z+1)(z-2)^2} = \dfrac{A}{z+1} + \dfrac{B}{z-2} + \dfrac{C}{(z-2)^2}$, where $A = B = C = 2.$

So that, $\dfrac{F(z)}{z} = \dfrac{2}{z+1} + \dfrac{2}{z-2} + \dfrac{2}{(z-2)^2}$

Hence, $F(z) = \dfrac{2z}{z+1} + \dfrac{2z}{z-2} + \dfrac{2z}{(z-2)^2}$

Inverting, we get

$$fn = 2(-1)^n + 2(2)^n + n.2^n$$

i.e, $fn = 2(-1)^n + (n+2)2^n$

Long Division Method

If $F(z)$ is expressed as a ratio of two polynomials, namely, $F(z)=g(z^{-1})/h(z^{-1})$, which can not be factorized, then divide the numerator by the denominator and the inverse transform can be taken term by term in the quotient.

Example:

Find the inverse $Z-$transform of $\frac{1+2z^{-1}}{1-z^{-1}}$ by long division method,

Let $F(z)=\frac{1+2z^{-1}}{1-z^{-1}}$

By actual division,

$$\begin{array}{r|l} & 1-3z^{-1}+3z^{-2}+3z^{-3} \\ \hline 1-z^{-1} & 1+2z^{-1} \\ & 1-z^{-1} \\ \hline & +3z^{-1} \\ & 3z^{-1}-3z^{-2} \\ \hline & +3z^{-2} \\ & 3z^{-2}-3z^{-3} \\ \hline & +3z^{-3} \\ & 3z^{-3}-3z^{-4} \\ \hline & +3z^{-4} \end{array}$$

Thus, $F(z)=1+3z-1+3z-2+3z-3+\ldots\ldots$

Now, Comparing the quotient with,

$$\sum_{n=0}^{\infty} f_n z^{-n} = f_0 + f_1 z^{-1} + f_2 z^{-2} + f_3 z^{-3} + \ldots\ldots$$

We get the sequence f_0 as $f_0=1,\ f_1=f_2=f_3=\ldots\ldots=3.$

$$\text{Hence } f_n = \begin{cases} 1, & \text{for } n=0 \\ 3, & \text{for } n \ge 1 \end{cases}$$

Example:

Find the inverse Z-transform of,

$$\frac{z}{z^2-3z+2}$$

By actual division,

$$\begin{array}{r|l} & z^{-1}+3z^{-1}+7z^{-3}+\dots\dots\dots \\ \hline 1-3z^{-1}+2z^{-2} & z^{-1} \\ & z^{-1}-3z^{-2}+2z^{-3} \\ \hline & \quad 3z^{-2}-2z^{-3} \\ & \quad 3z^{-2}-9z^{-3}+6z^{-4} \\ \hline & \qquad 7z^{-3}-6z^{-4} \\ & \qquad 7z^{-3}-21z^{-4}+14z^{-5} \\ \hline & \qquad +15z^{-4}-14z^{-5} \\ \hline \end{array}$$

$$\therefore F(z)=z^{-1}+3z^{-2}+7z^{-3}+\dots\dots\dots$$

Now comparing the quotient with,

$$\sum_{n=0}^{\infty} \mathrm{f}_n \mathrm{z}^{-n} = \mathrm{f}_0+\mathrm{f}_1\mathrm{z}^{-1}+\mathrm{f}_2\mathrm{z}^{-2}+\mathrm{f}_3\mathrm{z}^{-3}+\dots$$

We get the sequence f_n as $\mathrm{f}_0=0,\ \mathrm{f}_1=1, \mathrm{f}_2=3,\ \mathrm{f}_3=7\dots\dots\dots$

Hence, $\mathrm{f}_n = 2^n - 1, n = 0,1,2,3,\dots$

Solution of Difference Equations using Z-Transforms

First Order System

The first order discrete system is described by:

$$y[n]-ay[n-1]=x[n]$$

The impulse response $h[n]$ can be found by solving the following:

$$h[n]-ah[n-1]=\delta[n]$$

to be,

$$h[n]=a^n u[n]$$

Alternatively, we can take z-transform of the DE and get,

$$Y(z)-az^{-1}Y(z)=(1-az^{-1})Y(z)=X(z)$$

and the transfer function of the system (assumed causal),

$$H(z) = \frac{Y(z)}{X(z)} = \frac{1}{1 - az^{-1}} \qquad |z| > |a|$$

$H(z)$ has a zero at $z = 0$ and a pole $z_p = a$ and its ROC is the region $|z| > |a|$ outside the pole. If\ $|a| < 1$, then the unit circle $|z| = r = 1$ can be included in the ROC, the Fourier transform exists and the system is stable. The impulse response (unit sample response) of the system is,

$$h[n] = \mathcal{Z}^{-1}[H(z)] = a^n u[n]$$

Although h[n] seems to have a form different from the typical impulse response in continuous case $h(t) = e^{-t/\tau} u(t)$, they are essentially the same as $h(t)$ can be rewritten as,

$$h(t) = e^{-t/\tau} u(t) = (e^{-1/\tau})^t u(t) = a^t u(t)$$

Where,

$$a \stackrel{\Delta}{=} e^{-1/\tau}$$

Letting $z = e^{j\omega}$ in $H(z)$, we get the frequency response function of the system,

$$H(e^{j\omega}) = \frac{1}{1 - ae^{-j\omega}} = \frac{e^{j\omega}}{e^{j\omega} - a} = \frac{e^{j\omega} - 0}{e^{j\omega} - a} = \frac{u}{v}$$

where u and v are two vectors in z-plane defined as,

$$u \stackrel{\Delta}{=} e^{j\omega} - 0 = e^{j\omega}, \quad v \stackrel{\Delta}{=} e^{j\omega} - a$$

For any frequency $-\pi \le \omega \le \pi$ represented by a point $z = e^{j\omega}$ on the unit circle, the magnitude and phase angle of the frequency response function can be represented in the z-plane as,

$$|H(e^{j\omega})| = \frac{|u|}{|v|}$$

and

$$\angle H(e^{j\omega}) = \angle u - \angle v$$

which can be evaluated graphically in the z-plane as the frequency $\omega \backslash \omega$ changes in the range $-\pi \le \omega \le \pi$. If we assume $0 < a <$, then when $\omega = 0$, the denominator reaches its minimum of 1-a, and $|H(e^{j\omega})|$ is maximized to be 1/(1-a); and when $\omega = \pi$, the denominator reaches its maximum of |(-1-a)|=1+a, and $|H(e^{j\omega})|$ is minimized to be 1/(1+a). The phase angle of $(e^{j\omega}) = \angle u - \angle v$ is zero when $\omega = 0$ or $\omega = \pi$, and is negative for $0 < \omega < \pi$ and positive for $-\pi < \omega < 0$.

Second Order System

A second order discrete system is described by,

$$y[n]+a_1y[n-1]+a_2y[n-2]=x[n]$$

The Z-Transform of this DE is,

$$Y(z)+a_1z^{-1}Y(z)+a_2z^{-2}Y(z)=(1+a_1z^{-1}+a_2z^{-2})Y(z)=X(z)$$

and the transfer function is,

$$H(z)=\frac{Y(z)}{X(z)}=\frac{1}{1+a_1z^{-1}+a_2z^{-2}}=\frac{z^2}{z^2+a_1z+a_2}=\frac{z^2}{(z-z_{p_1})(z-z_{p_2})}$$

which has two repeated zeros z=0 and a pair of complex conjugate poles,

$$z_{p_1,p_2}=-\frac{a_1}{2}\pm\frac{1}{2}\sqrt{a_1^2-4a_2}=-\frac{a_1}{2}\pm\frac{j}{2}\sqrt{4a_2-a_1^2}$$

For convenience, we define,

$$a_1\overset{\Delta}{=}-2r\,cos\theta,\quad \text{and}\quad a_2\overset{\Delta}{=}r^2\quad (0\geq\theta\geq\pi)$$

and the transfer function becomes,

$$H(z)=\frac{z^2}{z^2-2r\,cos\theta z+r^2}$$

and the poles become,

$$z_{p_1,p_2}=r(cos\theta\pm j\,sin\theta)=r\,e^{\pm j\theta}$$

For the system to be causal and stable, we have to have $|z_p|=r|<1$. When $0<\theta<\pi$, $H(z)$ can be expanded,

$$H(z)=\frac{1}{(1-z_{p_1}z^{-1})(1-z_{p_2}z^{-1})}=\frac{1}{(1-re^{j\theta}z^{-1})(1-re^{-j\theta}z^{-1})}=\frac{A}{1-re^{j\theta}z^{-1}}+\frac{B}{1-re^{-j\theta}z^{-1}}$$

where,

$$A=\frac{e^{j\theta}}{2j\,sin\theta},\quad B=\frac{-e^{-j\theta}}{2j\,sin\theta}$$

The impulse response of the system can be found by inverse z-transform,

$$h[n]=[A(re^{j\theta})^n+B(re^{-j\theta})]u[n]=\frac{r^n}{2j\sin\theta}[e^{j\theta(n+1)}-e^{-j\theta(n+1)}]u[n]=r^n\frac{sin[(n+1)\theta]}{sin\theta}u[n]$$

When $\theta = 0$, $H(z)$ has two repeated poles $z_p = r$ and becomes,

$$H(z) = \frac{1}{(1 - rz^{-1})^2}$$

and the impulse response becomes,

$$h[n] = (n+1)r^n u[n]$$

When $\theta = \pi$, $H(z)$ has two repeated poles $z_p = -r$ and becomes,

$$H(z) = \frac{1}{(1 + rz^{-1})^2}$$

and the impulse response becomes,

$$h[n] = (n+1)(-r)^n u[n]$$

To graphically evaluate the behavior of the system as a function of frequency ω in the z-plane, we let $z = e^{j\omega}$ in $H(z)$ and get the frequency response function of the system,

$$H(e^{j\omega}) = \frac{1}{(1 - re^{j\theta}e^{-j\omega})(1 - re^{-j......j\theta})(e^{j\omega} - re^{-j\theta})} = \frac{u^2}{v_1 \, v_2}$$

where u, v^1 and v^2 are three vectors in z-plane defined as,

$$u \stackrel{\Delta}{=} e^{j\omega} - 0 = e^{j\omega}, \quad v_1 \stackrel{\Delta}{=} e^{j\omega} - re^{j\theta}, \quad v_2 \stackrel{\Delta}{=} e^{j\omega} - re^{-j\theta}$$

For any frequency $-\pi \le \omega \le \pi$ represented by a point $z = e^{j\omega}$ on the unit circle, the magnitude and phase angle of the frequency response function can be represented in the z-plane as,

$$|H(e^{j\omega})| = \frac{|u|^2}{|v_1||v_2|}$$

and

$$\angle H(e^{j\omega}) = 2\angle u - \angle v_1 - \angle v_2$$

When $\omega = \pm\pi$, v_1 and v_2 are maximized and thereby $|H(e^{j\omega})|$ is minimized. In particular, when r is close to 1, $|v_1| = 1 - |r|$ is minimized when $\omega = \theta$, i.e., $|H(e^{j\omega})|$ has a peak.

System Algebra and Block Diagram

Z-Transform converts time-domain operations such as difference and convolution into algebraic operations in z-domain. Moreover, the behavior of complex systems composed of a set of interconnected LTI systems can also be easily analyzed in z-domain. Some simple interconnections of LTI systems are listed below.

Parallel systems: If the system is composed of two LTI systems with $h_1[n]$ and $h_2[n]$ connected in parallel, its impulse response is,

$$h[n] = h_1[n] + h_2[n]$$

or in s-domain,

$$H(z) = H_1(z) + H_2(z)$$

Serial or cascade system: If the system is composed of two LTI systems with h1[n] and h2[n] connected in series, its impulse response is,

$$h[n] = h_1[n] * h_2[n] = h_2[n] * h_1[n]$$

or in s-domain,

$$H(z) = H_1(z)H_2(z) = H_2(z)H_1(z)$$

Feedback system: If the system is composed of an LTI system with $h_1[n]$ in a forward path and another LTI system $h_2[n]$ in a feedback path, its output $y[n]$ can be implicitly found in time domain,

$$y[n] = h_1[n] * e[n] = h_1[n] * [x[n] + h_2[n] * y[n]]$$

or in s-domain,

$$Y(z) = H_1(z)E(z) = H_1(z)\left[X(z) + H_2(z)Y(z)\right]$$

While it is difficult to solve the equation in time domain to find an explicit expression for output $y[n] = h[n] * x[n]$, it is easy to solve the algebraic equation in z-domain to find $y(z)$,

$$Y(z)[1 - H_1(z)H_2(z)] = H_1(z)X(z)$$

and the transfer function can be obtained,

$$H(z) = \frac{Y(z)}{X(z)} = \frac{H_1(z)}{1 - H_1(z)H_2(z)}$$

The feedback could be either positive or negative. For the latter, there will be a negative sign in front of $h_2(t)$ and $H_2(s)$ of the feedback path so that $e[n] = x[n] - h_2[n] * y[n]$ and

$$H(z) = \frac{Y(z)}{X(z)} = \frac{H_1(z)}{1 + H_1(z)H_2(z)}$$

Example: The transfer function of a first order LTI system,

$$y[n] - \frac{1}{4}y[n] = x[n]$$

is,

$$H(z)=\frac{1}{1-\frac{1}{4}z^{-1}}$$

Comparing this $H(z)$ with the transfer function of the feedback system, we see that a first order system can be represented as a feedback system with $H_1(z)=1$ in the forward path, and $H_2(z)$ for the product of 1/4 and z^{-1} (a delay element with input $x[n]$ and output $y[n]=x[n-1]$) in the negative feedback path. Example: Consider a second order system with transfer function,

$$H(z)=\frac{1}{1+\frac{1}{4}z^{-1}-\frac{1}{8}z^{-2}}=\frac{1}{(1+\frac{1}{2}z^{-1})(1-\frac{1}{4}z^{-1})}=\frac{2/3}{1+\frac{1}{2}z^{-1}}+\frac{1/3}{1-\frac{1}{4}z^{-1}}$$

These three expressions of this $H(z)$ correspond to three different block diagram representations of the system. The last two expressions are, respectively, the cascade and the parallel representations (same as the corresponding cases in Laplace transform), while the first one is the direct representation. We first consider a general 2nd order system,

$$Y(z)+asY(z)z^{-1}+bY(z)z^{-2}=X(z)$$

or,

$$Y(z)=X(z)-aY(z)z^{-1}-bY(z)z^{-2}$$

We see that Y(z) is the linear combination of the delayed versions of itself and the input X(z) which can be represented as a feedback system with two feedback paths of $-az^{-1}$ and $-bz^{-2}$. In this particular system, $a=-1/4$ an $b=1/8$. Example: A second order system with transfer function,

$$H(z)=\frac{c+dz^{-1}+ez^{-2}}{1+az^{-1}+bz^{-2}}=\frac{1}{1+az^{-1}+bz^{-2}}(c+dz^{-1}+ez^{-2})$$

This system can be represented as a cascade of two systems,

$$W(z)=H_1(z)X(z)=\frac{1}{1+az^{-1}+bz^{-2}}X(z)$$

and

$$Y(z)=H_2(z)W(z)=(c+dz^{-1}+ez^{-2})W(z)$$

The first system $H_1(z)$ can be implemented by two delay elements with proper feedback paths and the second system is a linear combination of $W(z)$, $W(z)z^{-1}$ and $W(z)z^{-2}$, all of which are available along the feedback path of the first system. The over all system can therefore be represented as shown. Obviously the block diagram of this example can be generalized to represent any system with a rational transfer function,

$$H(z) = \frac{\sum_{k=0}^{M} b_k z^{-k}}{\sum_{k=0}^{M} a_k z^{-k}} \quad (M \le N)$$

Unilateral Z-Transform

Unilateral Z-Transform of an arbitrary signal $x[n]$ is defined as:

$$UZ[x[n]] = X(z) \stackrel{\Delta}{=} \sum_{n=-\infty}^{\infty} x[n]u[n]z^{-n} = \sum_{n=0}^{\infty} x[n]z^{-n}$$

This definition always assumes $x[n] = 0$ for $t < 0$. When the unilateral Z-Transform is applied to find the transfer function $H(z) = UZ[h[n]]$ of an LTI system, it is always assumed to be causal, and the ROC is always the exterior of a circle. By definition, the unilateral z-transform of any signal $x[n] = x[n]u[n]$ is identical to its bilateral Laplace transform. However, when $x[n] \ne x[n]u[n]$, the two z-transforms are different. Some of the properties of the unilateral Z-Transform different from the bilateral Z-Transform are listed below.

Time Advance:

$$UZ[x[n+1]] = \sum_{n=0}^{\infty} x[n+1]z^{-n} = z\sum_{n=0}^{\infty} x[m]z^{-m} = z[\sum_{m=0}^{\infty} x[m]z^{-m} - x[0]] = zX(z) - zx[0]$$

where $m = n+1$

Time Delay:

$$UZ[x[n-1]] = \sum_{n=0}^{\infty} x[n-1]z^{-n} = z^{-1}\sum_{m=-1}^{\infty} x[m]z^{-m}$$

$$z^{-1}[\sum_{m=0}^{\infty} x[m]z^{-m} + zx[-1]] = z^{-1}X(z) + x[-1]$$

where $m = n-1$. Similarly, we have,

$$UZ[x[n-2]] = \sum_{n=0}^{\infty} x[n-2]z^{-n} = z^{-2}\sum_{m=-2}^{\infty} x[m]z^{-m}$$

$$= z^{-2}[\sum_{m=0}^{\infty} x[m]z^{-m} + zx[-1] + z^{2}x[-2]]$$

$$= z^{-2}X(z) + x[-1]z^{-1} + x[-2]$$

where $m = n-2$. In general, we have:

$$UZ[x[n-n_0]] = z^{-n_0}X(z) + \sum_{k=0}^{n_0-1} z^{-k}x[k-n_0]$$

Convolution:

$$UZ[x[n]*y[n]] = X(z)Y(z)$$

If both $x[n]$ and $y[n]$ are causal, i.e., $x[n] = y[n] = 0$ for $n < 0$, the unilateral and bilateral z-transforms are identical.

Time Difference:

$$UZ[x[n]-x[n-1]] = (1-z^{-1})X(z) \qquad ROC = R_x$$

Proof:

$$UZ[x[n]-x[n-1]] = X(z) - z^{-1}X(z) - x[-1] = (1-z^{-1})X(z) - x[-1]$$

Time Accumulation:

$$UZ[\sum_{k=0}^{n} x[k]] = \frac{1}{1-z^{-1}} X(z)$$

Initial Value Theorem If $x[n] = x[n]u[n]$, i.e., $x[n] = 0$ for $n < 0$, then

$$x[0] = \lim_{z\to\infty} X(z)$$

Proof:

$$\lim_{z\to\infty} X(z) = \lim_{z\to\infty} [\sum_{n=0}^{\infty} x[n]z^{-n}] = x[0]$$

All terms with n>0 become zero as $z^{-n} = 1/z^n \to 0$ as $z \to \infty$, except the first one which is always $x[0]$.

Final Value Theorem If $x[n] = x[n]u[n]$, i.e., $x[n] = 0$ for $n < 0$, then

$$\lim_{n\to\infty} x[n] = \lim_{z\to 1} (1-z^{-1})X(z)$$

Proof:

$$Z[x[n]-x[n-1]] = X(z) - X(z)z^{-1} = \sum_{m=-2}^{\infty} [x[n]-x[n-1]]z^{-n}$$

i.e.

$$(1-z^{-1})X(z) = \lim_{N\to\infty} \sum_{n=0}^{N} [x[n]-x[n-1]]z^{-n}$$

Letting $z \to 1$ in the above, we get,

$$\lim_{z\to 1}(1-z^{-1})X(z) = \lim_{N\to\infty}\sum_{n=0}^{N}[x[n]-x[n-1]]$$

$$= \lim_{N\to\infty}\{\sum_{n=0}^{N-1}[x[n]-x[n]]+x[N]-x[-1]\} = \lim_{N\to\infty} x[N]$$

as $x[-1]=0$.

Example:

$$x[n] = a^{n+1}u[n+1]$$

This signal is right sided starting at n=-1 (i.e., $x[n] \neq x[n]u[n]$). By definition, the bilateral z-transform of $x[n]$ is,

$$Z[x[n]] = \sum_{n=-1}^{\infty} a^{n+1}z^{-n} = z + a\sum_{n=0}^{\infty} a^n z^{-n} = z + \frac{a}{1-az^{-1}} = \frac{z}{1-az^{-1}}$$

It was assumed that $|z|>|a|$. The unilateral z-transform of this signal is,

$$UZ[x[n]] = \sum_{n=0}^{\infty} a^{n+1}z^{-n} = a\sum_{n=0}^{\infty} a^n z^{-n} = \frac{a}{1-az^{-1}}$$

If we assume zero initial condition $y[-1]=0$,

Solving LCCDEs by Unilateral Z-Transform

Due to its time delay property, the unilateral z-transform is a powerful tool for solving LCCDEs with arbitrary initial conditions. Example: A system is described by this LCCDE,

$$y[n]+3y[n-1] = x[n] = \alpha u[n]$$

Taking unilateral z-transform of the DE, we get

$$Y(z)+3Y(z)z^{-1}+3y[-1] = X(z) = \frac{\alpha}{1-z^{-1}}$$

The particular (zero-state) solution If the system is initially at rest, i.e., $y[-1]=0$, the above equation can be solved for the output $Y(z)$ to get,

$$Y(z) = H(z)X(z) = \frac{1}{1+3z^{-1}}\frac{\alpha}{1-z^{-1}} = \frac{3\alpha/4}{1+3z^{-1}} + \frac{\alpha/4}{1-z^{-1}}$$

where $H(z)=1/(1+3z^{-1})$ is the system's transfer function. In time domain this is the particular (or zero-state) solution (caused by the input with zero initial condition):

$$y_p[n]=\alpha[\frac{1}{4}+\frac{3}{4}(-3)^n]u[n]$$

The homogeneous (zero-input) solution When the initial condition is nonzero,

$$y[-1]=\beta$$

but the input is zero $x[n]=0$, , the z-transform of the difference equation becomes,

$$Y(z)+3Y(z)z^{-1}+3\beta=0$$

which can be solved for $Y(z)$

$$Y(z)=\frac{-3\beta}{1+3z^{-1}}$$

In time domain, this is the homogeneous (or zero-input) solution (caused by the initial condition with zero input):

$$y_h[n]=-3\beta(-3)^n u[n]$$

When neither $y[-1]$ nor $x[n]$ is zero, we have:

$$Y(z)+3Y(z)z^{-1}+3\beta=X(z)=\frac{\alpha}{1-z^{-1}}$$

Solving this algebraic equation in z-domain for $Y(z)$ we get

$$Y(z)=\frac{\alpha}{(1+3z^{-1})(1-z^{-1})}-\frac{3\beta}{1+3z^{-1}}$$

The first term is the particular solution caused by the input alone and the second term is the homogeneous solution caused by the initial condition alone. The $Y(z)$ can be further written as:

$$Y(z)=\frac{1}{1+3z^{-1}}(\frac{3}{4}\alpha-3\beta)+\frac{\alpha}{4}\frac{1}{1-z^{-1}}$$

and in time domain, we have the general solution,

$$y_g[n]=[(\frac{3}{4}\alpha-3\beta)(-3)^n+\frac{\alpha}{4}]u[n]=y_h[n]+y_p[n]$$

which is the sum of both the homogeneous and particular solutions. Note that bilateral Z-Transform can also be used to solve LCCDEs. However, as bilateral Z-Transform does not take initial condition into account, it is always implicitly assumed that the system is initially at rest.

Analysis of LTI Systems by Z-Transform

Due to its convolution property, the z-transform is a powerful tool to analyze LTI systems:

$$y[n] = h[n] * x[n] \xrightarrow{Z} Y(z) = H(z)X(z)$$

when the input is the eigenfunction of all LTI system, i.e., $x[n] = e^{sn} = z^n$, the operation on this input by the system can be found by multiplying the system's eigenvalue $H(z)$ to the input:

$$y[n] = O[z^n] = h[n] * z^n = H(z)z^n$$

Causal LTI Systems

An LTI system is causal if its output $y[n]$ depends only on the current and past input $x[n]$ (but not the future). Assuming the system is initially at rest with zero output $y[n]\big|_{n<0} = 0$, then its response $y[n] = h[n]$ to an impulse $x[n] = \delta[n]$ at $n = 0$ is at rest for $n < 0$ i.e., $h[n] = h[n]u[n]$. Its response to a general input $x(t)$ is:

$$y[n] = h[n] * x[n] = \sum_{m=-\infty}^{\infty} h[m]x[n-m] = \sum_{m=0}^{\infty} h[m]x[n-m]$$

Due to the properties of the ROC, we know that:

If an LTI system is causal (with a right sided impulse response function $h[n] = 0$ for, $n < 0$) then the ROC of its transfer function $H(z)$ is the exterior of a circle including infinity. In particular, when $H(z)$ is rational, then the system is causal if and only if its ROC is the exterior of a circle outside the out-most pole, and the order of numerator is no greater than the order of the denominator.

Note the requirement for the orders of the numerator and denominator guarantees the existence of $H(z)$ even when $z = \infty$.

Stable LTI Systems

An LTI system is stable if its response to any bounded input is also bounded for all n:

$$\text{if } |x[n]| < B_x \quad \text{then} \quad |y[n]| < \infty$$

As the output and input of an LTI is related by convolution, we have:

$$y[n] = h[n] * x[n] = \sum_{m=-\infty}^{\infty} h[m]x[n-m] < \infty$$

$$|y[n]|=|\sum_{m=-\infty}^{\infty} h[m]x[n-m]| \le \sum_{m=-\infty}^{\infty} |h[m]||x[n-m]| < B_x \sum_{m=-\infty}^{\infty} |h[m]| < \infty$$

which obviously requires:

$$\sum_{m=-\infty}^{\infty} |h[m]| < \infty$$

In other words, if the impulse response function $h[m]$ of an LTI system is absolutely integrable, then the system is stable. We can show that this condition is also necessary, i.e., all stable LTI systems' impulse response functions are absolutely integrable.

An LTI system is stable if and only if its impulse response is absolutely summable, i.e., the frequency response function $H(e^{j\omega})$ exits, i.e. the ROC of its transfer function $H(z)$ includes the unit circle $|z|=1$.

$$H(z)\Big|_{z=e^{j\omega}} = H(e^{j\omega}) = F[h[n]]$$

Causal and Stable LTI Systems

From the two properties above, we also see that:

A causal LTI system with a rational transfer function $H(z)$ is stable if and only if all poles of $H(z)$ are inside the unit circle of the z-plane, i.e., the magnitudes of all poles are smaller than 1.

$$|Re[z_p]| < 1 \qquad (\text{for all } z_p)$$

Example: The transfer function of an LTI is,

$$H(z) = \frac{1}{1-az^{-1}}$$

As shown before, without specifying the ROC, this $H(z)$ could be the z-transform of one of the two possible time signals $h[n]$.

- If ROC is $|z|>|a|$, the system $h[n]=-a^n u[n]$ is causal.
  - If $|a|<1$, i.e., unit circle can be included in ROC, the system is stable;
  - If $|a|>1$, i.e., unit circle cannot be included in ROC, the system is unstable;
- If ROC is $|z|<|a|$, the system $h[n]=-a\ u[-n-1]$ is anti-causal.
  - If $|a|<1$, i.e., unit circle cannot be included in ROC, the system is unstable;
  - If $|a|>1$, i.e., unit circle can be included in ROC, the system is stable;

	$\|a\|<1$	$\|a\|>1$
$\|z\|>\|a\|$	$e^{j\omega}$ inside ROC	$e^{j\omega}$ outside ROC,

$h[n]=-a^n u[n]$	causal, stable	causal, unstable
$\|z\|<\|a\|$	$e^{j\omega}$ outside ROC	$e^{j\omega}$ inside ROC,
$h[n]=-a^n u[-n-1]$	anti-causal, unstable	anti-causal, stable

References

- Introduction-z-transform-examples: myclassbook.org, Retrieved 19 April, 2019
- Node6, z-transform: fourier.eng.hmc.edu, Retrieved 5 February, 2019
- Inverse-z-transforms_6511: brainkart.com, Retrieved 26 July, 2019
- Solution-difference-equations-using-z-transform, classle: classle.net, Retrieved 21 May, 2019
- Node6, z-transform, lectures: fourier.eng.hmc.edu, Retrieved 8 January, 2019

6

Digital Filter and its Types

The development of a signal flow diagram which describes the filter in terms of operations on sample sequences is termed as digital filter realization. A few of the different types of filters are low-pass filters, high-pass filters, all-pass filters, anti-aliasing filters, etc. All these types of digital filter realizations and filters have been carefully analyzed in this chapter.

Filter

In signal processing, a filter is a device or process that removes some unwanted components or features from a signal. Filtering is a class of signal processing, the defining feature of filters being the complete or partial suppression of some aspect of the signal.

Most often, this means removing some frequencies or frequency bands. However, filters do not exclusively act in the frequency domain; especially in the field of image processing many other targets for filtering exist. Correlations can be removed for certain frequency components and not for others without having to act in the frequency domain. Filters are widely used in electronics and telecommunication, in radio, television, audio recording, radar, control systems, music synthesis, image processing, and computer graphics.

There are many different bases of classifying filters and these overlap in many different ways; there is no simple hierarchical classification. Filters may be:

- Non-linear or linear.
- Time-variant or time-invariant, also known as shift invariance. If the filter operates in a spatial domain then the characterization is space invariance.
- Causal or not-causal: A filter is non-causal if its present output depends on future input. Filters processing time-domain signals in real time must be causal, but not filters acting on spatial domain signals or deferred-time processing of time-domain signals.
- Analog or digital.
- Discrete-time (sampled) or continuous-time.

- Passive or active type of continuous-time filter.
- Infinite impulse response (IIR) or finite impulse response (FIR) type of discrete-time or digital filter.

Technologies

Filters can be built in a number of different technologies. The same transfer function can be realised in several different ways, that is the mathematical properties of the filter are the same but the physical properties are quite different. Often the components in different technologies are directly analogous to each other and fulfill the same role in their respective filters. For instance, the resistors, inductors and capacitors of electronics correspond respectively to dampers, masses and springs in mechanics. Likewise, there are corresponding components in distributed-element filters.

- Electronic filters were originally entirely passive consisting of resistance, inductance and capacitance. Active technology makes design easier and opens up new possibilities in filter specifications.
- Digital filters operate on signals represented in digital form. The essence of a digital filter is that it directly implements a mathematical algorithm, corresponding to the desired filter transfer function, in its programming or microcode.
- Mechanical filters are built out of mechanical components. In the vast majority of cases they are used to process an electronic signal and transducers are provided to convert this to and from a mechanical vibration. However, examples do exist of filters that have been designed for operation entirely in the mechanical domain.
- Distributed-element filters are constructed out of components made from small pieces of transmission line or other distributed elements. There are structures in distributed-element filters that directly correspond to the lumped elements of electronic filters, and others that are unique to this class of technology.
- Waveguide filters consist of waveguide components or components inserted in the waveguide. Waveguides are a class of transmission line and many structures of distributed-element filters, for instance the stub, can also be implemented in waveguides.
- Optical filters were originally developed for purposes other than signal processing such as lighting and photography. With the rise of optical fiber technology, however, optical filters increasingly find signal processing applications and signal processing filter terminology, such as longpass and shortpass, are entering the field.
- Transversal filter, or delay line filter, works by summing copies of the input after various time delays. This can be implemented with various technologies including analog delay lines, active circuitry, CCD delay lines, or entirely in the digital domain.

Digital Filters

Digital signal processing allows the inexpensive construction of a wide variety of filters. The signal is sampled and an analog-to-digital converter turns the signal into a stream of numbers. A computer program running on a CPU or a specialized DSP (or less often running on a hardware implementation of the algorithm) calculates an output number stream. This output can be converted to a signal by passing it through a digital-to-analog converter. There are problems with noise introduced by the conversions, but these can be controlled and limited for many useful filters. Due to the sampling involved, the input signal must be of limited frequency content or aliasing will occur.

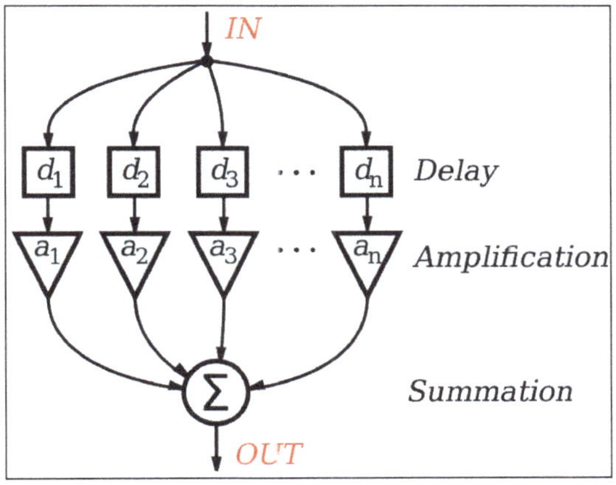

A general finite impulse response filter with n stages, each with an independent delay, d_i and amplification gain, a_i.

Quartz Filters and Piezoelectrics

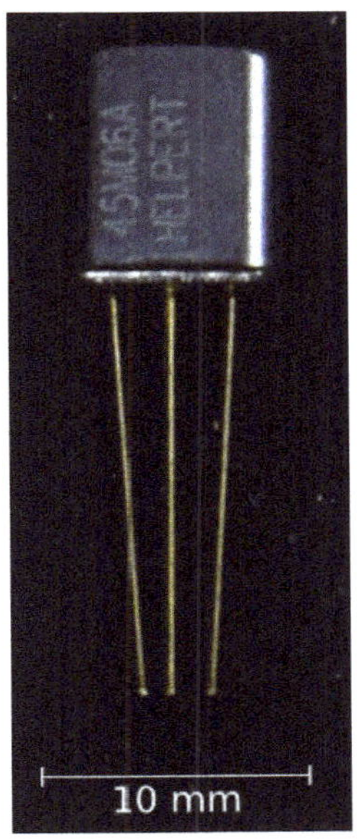

Crystal filter with a center frequency of 45 MHz and a bandwidth B_{3dB} of 12 kHz.

In the late 1930s, engineers realized that small mechanical systems made of rigid materials such as quartz would acoustically resonate at radio frequencies, i.e. from audible frequencies (sound) up to several hundred megahertz. Some early resonators were made of steel, but quartz quickly became favored. The biggest advantage of quartz is that it is piezoelectric. This means that quartz resonators can directly convert their own mechanical motion into electrical signals. Quartz also has a very low coefficient of thermal expansion which means that quartz resonators can produce stable frequencies over a wide temperature range. Quartz crystal filters have much higher quality factors than LCR filters. When higher stabilities are required, the crystals and their driving circuits may be mounted in a "crystal oven" to control the temperature. For very narrow band filters, sometimes several crystals are operated in series.

A large number of crystals can be collapsed into a single component, by mounting comb-shaped evaporations of metal on a quartz crystal. In this scheme, a "tapped delay line" reinforces the desired frequencies as the sound waves flow across the surface of the quartz crystal. The tapped delay line has become a general scheme of making high-Q filters in many different ways.

SAW Filters

SAW (surface acoustic wave) filters are electromechanical devices commonly used in radio frequency applications. Electrical signals are converted to a mechanical wave in a device constructed of a piezoelectric crystal or ceramic; this wave is delayed as it propagates across the device, before being converted back to an electrical signal by further electrodes. The delayed outputs are recombined to produce a direct analog implementation of a finite impulse response filter. This hybrid filtering technique is also found in an analog sampled filter. SAW filters are limited to frequencies up to 3 GHz. The filters were developed by Professor Ted Paige and others.

BAW Filters

BAW (bulk acoustic wave) filters are electromechanical devices. BAW filters can implement ladder or lattice filters. BAW filters typically operate at frequencies from around 2 to around 16 GHz, and may be smaller or thinner than equivalent SAW filters. Two main variants of BAW filters are making their way into devices: thin-film bulk acoustic resonator or FBAR and solid mounted bulk acoustic resonators.

Garnet Filters

Another method of filtering, at microwave frequencies from 800 MHz to about 5 GHz, is to use a synthetic single crystal yttrium iron garnet sphere made of a chemical combination of yttrium and iron (YIGF, or yttrium iron garnet filter). The garnet sits on a strip of metal driven by a transistor, and a small loop antenna touches the top of the sphere. An electromagnet changes the frequency that the garnet will pass. The advantage of this method is that the garnet can be tuned over a very wide frequency by varying the strength of the magnetic field.

Atomic Filters

For even higher frequencies and greater precision, the vibrations of atoms must be used. Atomic clocks use caesium masers as ultra-high Q filters to stabilize their primary oscillators. Another method, used at high, fixed frequencies with very weak radio signals, is to use a ruby maser tapped delay line.

The Transfer Function

The transfer function of a filter is most often defined in the domain of the complex frequencies. The back and forth passage to/from this domain is operated by the Laplace transform and its inverse (therefore, here below, the term "input signal" shall be understood as "the Laplace transform of" (the time representation of) the input signal, and so on).

The transfer function $H(s)$ of a filter is the ratio of the output signal $Y(s$ to that of the input signal $X(s)$ as a function of the complex frequency:

$$H(s)=\frac{Y(s)}{X(s)}$$

with $s=\sigma+j\omega$.

The transfer function of all linear time-invariant filters generally share certain characteristics:

- For filters which are constructed of discrete components, their transfer function must be the ratio of two polynomials in s, i.e. a rational function of s. The order of the transfer function will be the highest power of s encountered in either the numerator or the denominator.
- The polynomials of the transfer function will all have real coefficients. Therefore, the poles and zeroes of the transfer function will either be real or occur in complex conjugate pairs.
- Since the filters are assumed to be stable, the real part of all poles (i.e. zeroes of the denominator) will be negative, i.e. they will lie in the left half-plane in complex frequency space.

Distributed-element filters do not, in general, produce rational functions, but can approximate them.

The proper construction of a transfer function involves the Laplace transform, and therefore it is needed to assume null initial conditions, because

$$\mathcal{L}\left\{\frac{df}{dt}\right\}=s\cdot\mathcal{L}\{f(t)\}-f(0),$$

And when f(0)=0 we can get rid of the constants and use the usual expression,

$$\mathcal{L}\left\{\frac{df}{dt}\right\}=s\cdot\mathcal{L}\{f(t)\}$$

An alternative to transfer functions is to give the behavior of the filter as a convolution. The convolution theorem, which holds for Laplace transforms, guarantees equivalence with transfer functions.

Classification

Filters may be specified by family and bandform. A filter's family is specified by the approximating polynomial used and each leads to certain characteristics of the transfer function of the filter. Some common filter families and their particular characteristics are:

- Butterworth filter – no gain ripple in pass band and stop band, slow cutoff.

- Chebyshev filter (Type I) – no gain ripple in stop band, moderate cutoff.
- Chebyshev filter (Type II) – no gain ripple in pass band, moderate cutoff.
- Bessel filter – no group delay ripple, no gain ripple in both bands, slow gain cutoff.
- Elliptic filter – gain ripple in pass and stop band, fast cutoff.
- Optimum "L" filter.
- Gaussian filter – no ripple in response to step function.
- Hourglass filter.
- Raised-cosine filter.

Each family of filters can be specified to a particular order. The higher the order, the more the filter will approach the "ideal" filter; but also the longer the impulse response is and the longer the latency will be. An ideal filter has full transmission in the pass band, complete attenuation in the stop band, and an abrupt transition between the two bands, but this filter has infinite order (i.e., the response cannot be expressed as a linear differential equation with a finite sum) and infinite latency (i.e., its compact support in the Fourier transform forces its time response to be ever lasting).

Here is an image comparing Butterworth, Chebyshev, and elliptic filters. The filters in this illustration are all fifth-order low-pass filters. The particular implementation – analog or digital, passive or active – makes no difference; their output would be the same.

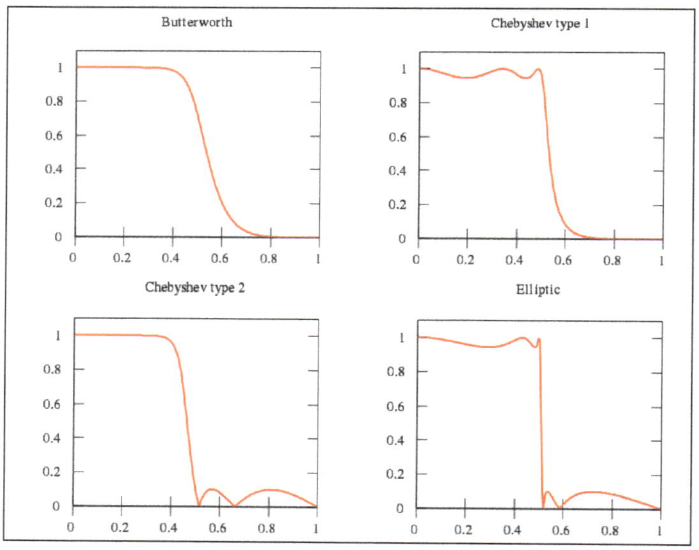

As is clear from the image, elliptic filters are sharper than all the others, but they show ripples on the whole bandwidth.

Any family can be used to implement a particular bandform of which frequencies are transmitted, and which, outside the passband, are more or less attenuated. The transfer function completely specifies the behavior of a linear filter, but not the particular technology used to implement it. In other words, there are a number of different ways of achieving a particular transfer function when designing a circuit. A particular bandform of filter can be obtained by transformation of a prototype filter of that family.

Impedance Matching

Impedance matching structures invariably take on the form of a filter, that is, a network of non-dissipative elements. For instance, in a passive electronics implementation, it would likely take the form of a ladder topology of inductors and capacitors. The design of matching networks shares much in common with filters and the design invariably will have a filtering action as an incidental consequence. Although the prime purpose of a matching network is not to filter, it is often the case that both functions are combined in the same circuit. The need for impedance matching does not arise while signals are in the digital domain.

Similar comments can be made regarding power dividers and directional couplers. When implemented in a distributed-element format, these devices can take the form of a distributed-element filter. There are four ports to be matched and widening the bandwidth requires filter-like structures to achieve this. The inverse is also true: distributed-element filters can take the form of coupled lines.

Low-pass Filter

A low-pass filter (LPF) is a filter that passes signals with a frequency lower than a selected cutoff frequency and attenuates signals with frequencies higher than the cutoff frequency. The exact frequency response of the filter depends on the filter design. The filter is sometimes called a high-cut filter, or treble-cut filter in audio applications. A low-pass filter is the complement of a high-pass filter.

In the optical domain, high-pass and low-pass have the opposite meanings, with a “high-pass” filter (more commonly “long-pass”) passing only *longer* wavelengths (lower frequencies), and vice-versa for “low-pass” (more commonly “short-pass”).

Low-pass filters exist in many different forms, including electronic circuits such as a hiss filter used in audio, anti-aliasing filters for conditioning signals prior to analog-to-digital conversion, digital filters for smoothing sets of data, acoustic barriers, blurring of images, and so on. The moving average operation used in fields such as finance is a particular kind of low-pass filter, and can be analyzed with the same signal processing techniques as are used for other low-pass filters. Low-pass filters provide a smoother form of a signal, removing the short-term fluctuations and leaving the longer-term trend.

Filter designers will often use the low-pass form as a prototype filter. That is, a filter with unity bandwidth and impedance. The desired filter is obtained from the prototype by scaling for the desired bandwidth and impedance and transforming into the desired bandform (that is low-pass, high-pass, band-pass or band-stop).

Examples of low-pass filters occur in acoustics, optics and electronics.

A stiff physical barrier tends to reflect higher sound frequencies, and so acts as an acoustic low-pass filter for transmitting sound. When music is playing in another room, the low notes are easily heard, while the high notes are attenuated.

An optical filter with the same function can correctly be called a low-pass filter, but conventionally is called a *longpass* filter (low frequency is long wavelength), to avoid confusion.

In an electronic low-pass RC filter for voltage signals, high frequencies in the input signal are attenuated, but the filter has little attenuation below the cutoff frequency determined by its RC time constant. For current signals, a similar circuit, using a resistor and capacitor in parallel, works in a similar manner.

Electronic low-pass filters are used on inputs to subwoofers and other types of loudspeakers, to block high pitches that they can't efficiently reproduce. Radio transmitters use low-pass filters to block harmonic emissions that might interfere with other communications. The tone knob on many electric guitars is a low-pass filter used to reduce the amount of treble in the sound. An integrator is another time constant low-pass filter.

Telephone lines fitted with DSL splitters use low-pass and high-pass filters to separate DSL and POTS signals sharing the same pair of wires.

Low-pass filters also play a significant role in the sculpting of sound created by analogue and virtual analogue synthesisers.

A low-pass filter is used as an anti-aliasing filter prior to sampling and for reconstruction in digital-to-analog conversion.

Ideal and Real Filters

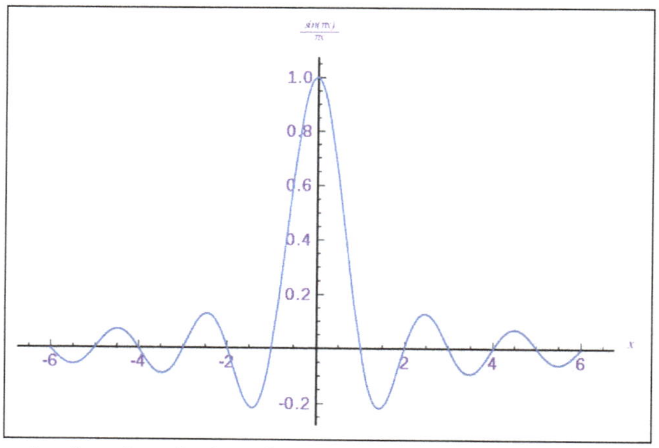

The sinc function, the impulse response of an ideal low-pass filter.

An ideal low-pass filter completely eliminates all frequencies above the cutoff frequency while passing those below unchanged; its frequency response is a rectangular function and is a brick-wall filter. The transition region present in practical filters does not exist in an ideal filter. An ideal low-pass filter can be realized mathematically (theoretically) by multiplying a signal by the rectangular function in the frequency domain or, equivalently, convolution with its impulse response, a sinc function, in the time domain.

However, the ideal filter is impossible to realize without also having signals of infinite extent in time, and so generally needs to be approximated for real ongoing signals, because the sinc function's support region extends to all past and future times. The filter would therefore need to have infinite delay, or knowledge of the infinite future and past, in order to perform the convolution. It is effectively realizable for pre-recorded digital signals by assuming extensions of zero into the past and future, or more typically by making the signal repetitive and using Fourier analysis.

Real filters for real-time applications approximate the ideal filter by truncating and windowing the infinite impulse response to make a finite impulse response; applying that filter requires delaying the signal for a moderate period of time, allowing the computation to "see" a little bit into the future. This delay is manifested as phase shift. Greater accuracy in approximation requires a longer delay.

An ideal low-pass filter results in ringing artifacts via the Gibbs phenomenon. These can be reduced or worsened by choice of windowing function, and the design and choice of real filters involves understanding and minimizing these artifacts. For example, "simple truncation [of sinc] causes severe ringing artifacts," in signal reconstruction, and to reduce these artifacts one uses window functions "which drop off more smoothly at the edges."

The Whittaker–Shannon interpolation formula describes how to use a perfect low-pass filter to reconstruct a continuous signal from a sampled digital signal. Real digital-to-analog converters use real filter approximations.

Discrete-time Realization

For another method of conversion from continuous- to discrete-time,

Many digital filters are designed to give low-pass characteristics. Both infinite impulse response and finite impulse response low pass filters as well as filters using Fourier transforms are widely used.

Simple Infinite Impulse Response Filter

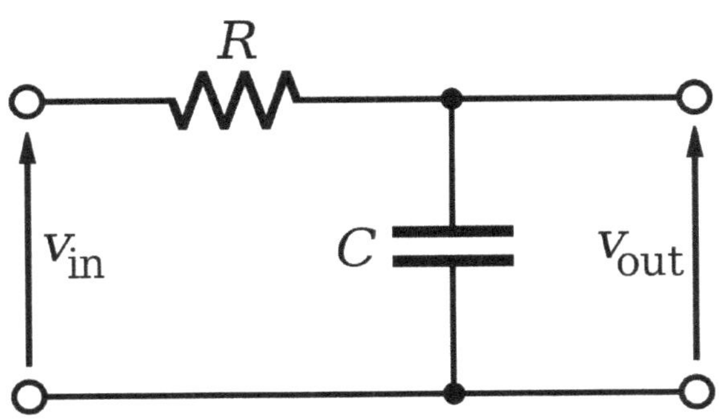

A simple low-pass RC filter

The effect of an infinite impulse response low-pass filter can be simulated on a computer by analyzing an RC filter's behavior in the time domain, and then discretizing the model.

From the circuit diagram to the right, according to Kirchhoff's Laws and the definition of capacitance:

$$v_{\text{in}}(t) - v_{\text{out}}(t) = R\, i(t)$$

$$Q_c(t) = Cv_{\text{out}}(t)$$

$$i(t) = \frac{\mathrm{d}Q_c}{\mathrm{d}t}$$

where $Q_c(t)$ is the charge stored in the capacitor at time t. Substituting equation Q into equation I gives $i(t) = C\frac{\mathrm{d}v_{\text{out}}}{\mathrm{d}t}$, which can be substituted into equation V so that:

$$v_{\text{in}}(t) - v_{\text{out}}(t) = RC\frac{\mathrm{d}v_{\text{out}}}{\mathrm{d}t}$$

This equation can be discretized. For simplicity, assume that samples of the input and output are taken at evenly spaced points in time separated by Δ_T time. Let the samples of v_{in} be represented by the sequence $(x_1, x_2, \ldots, x_n)$, and let v_{out} be represented by the sequence $(y_1, y_2, \ldots, y_n)$, which correspond to the same points in time. Making these substitutions:

$$x_i - y_i = RC\frac{y_i - y_{i-1}}{\Delta_T}$$

And rearranging terms gives the recurrence relation,

$$y_i = \overbrace{x_i\left(\frac{\Delta_T}{RC + \Delta_T}\right)}^{\text{Input contribution}} + \overbrace{y_{i-1}\left(\frac{RC}{RC + \Delta_T}\right)}^{\text{Inertia from previous output}}.$$

That is, this discrete-time implementation of a simple RC low-pass filter is the exponentially weighted moving average,

$$y_i = \alpha x_i + (1 - \alpha)y_{i-1} \qquad \text{where} \qquad \alpha := \frac{\Delta_T}{RC + \Delta_T}$$

By definition, the *smoothing factor* $0 \le \alpha \le 1$. The expression for α yields the equivalent time constant RC in terms of the sampling period Δ_T and smoothing factor:

$$RC = \Delta_T\left(\frac{1 - \alpha}{\alpha}\right)$$

Recalling that

$$f_c = \frac{1}{2\pi RC} \quad \text{so} \quad RC = \frac{1}{2\pi f_c}$$

then α and f_c are related by,

$$\alpha = \frac{2\pi\Delta_T f_c}{2\pi\Delta_T f_c + 1}$$

and

$$f_c = \frac{\alpha}{(1-\alpha)2\pi\Delta_T}$$

If $\alpha = 0.5$, then the RC time constant is equal to the sampling period. If $\alpha \ll 0.5$, then RC is significantly larger than the sampling interval, and $\Delta_T \approx \alpha RC$.

The filter recurrence relation provides a way to determine the output samples in terms of the input samples and the preceding output. The following pseudocode algorithm simulates the effect of a low-pass filter on a series of digital samples:

```
// Return RC low-pass filter output samples, given input samples,
// time interval dt, and time constant RC
function lowpass(real[0..n] x, real dt, real RC)
 var real[0..n] y
 var real α := dt / (RC + dt)
    y := α * x
 for i from 1 to n
    y[i] := α * x[i] + (1-α) * y[i-1]
return y
```

The loop that calculates each of the n outputs can be refactored into the equivalent:

```
for i from 1 to n
   y[i] := y[i-1] + α * (x[i] - y[i-1])
```

That is, the change from one filter output to the next is proportional to the difference between the previous output and the next input. This exponential smoothing property matches the exponential decay seen in the continuous-time system. As expected, as the time constant RC increases, the discrete-time smoothing parameter α decreases, and the output samples $(y_1, y_2, \ldots, y_n)$ respond more slowly to a change in the input samples $(x_1, x_2, \ldots, x_n)$; the system has more *inertia*. This filter is an infinite-impulse-response (IIR) single-pole low-pass filter.

Finite Impulse Response

Finite-impulse-response filters can be built that approximate to the sinc function time-domain response of an ideal sharp-cutoff low-pass filter. For minimum distortion the finite impulse response filter has an unbounded number of coefficients operating on an unbounded signal. In practice, the time-domain response must be time truncated and is often of a simplified shape; in the simplest case, a running average can be used, giving a square time response.

Fourier Transform

For non-realtime filtering, to achieve a low pass filter, the entire signal is usually taken as a looped signal, the Fourier transform is taken, filtered in the frequency domain, followed by an inverse Fourier transform. Only O(n log(n)) operations are required compared to O(n^2) for the time domain filtering algorithm.

This can also sometimes be done in real-time, where the signal is delayed long enough to perform the Fourier transformation on shorter, overlapping blocks.

Continuous-time Realization

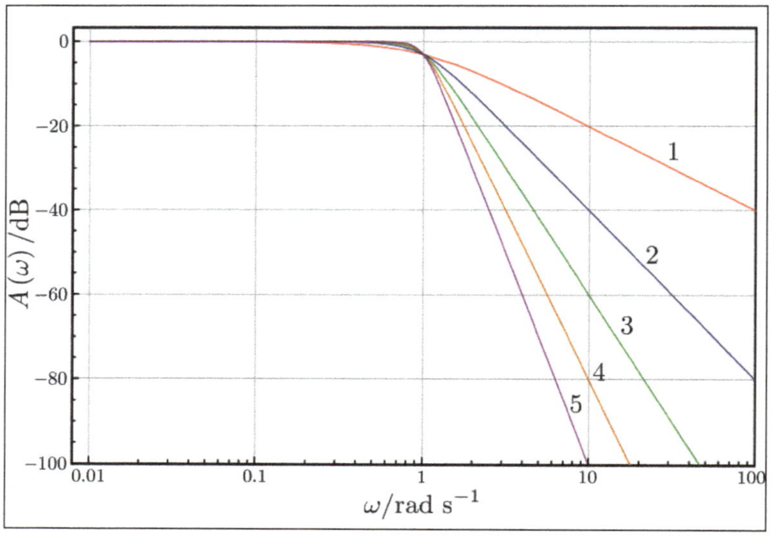

In the figure above, gain-magnitude frequency response of a first-order (one-pole) low-pass filter. *Power gain* is shown in decibels (i.e., a 3 dB decline reflects an additional half-power attenuation). Angular frequency is shown on a logarithmic scale in units of radians per second.

There are many different types of filter circuits, with different responses to changing frequency. The frequency response of a filter is generally represented using a Bode plot, and the filter is characterized by its cutoff frequency and rate of frequency rolloff. In all cases, at the *cutoff frequency,* the filter attenuates the input power by half or 3 dB. So the order of the filter determines the amount of additional attenuation for frequencies higher than the cutoff frequency.

- A first-order filter, for example, reduces the signal amplitude by half (so power reduces by a factor of 4, or 6 dB), every time the frequency doubles (goes up one octave); more precisely, the power rolloff approaches 20 dB per decade in the limit of high frequency. The magnitude Bode plot for a first-order filter looks like a horizontal line below the cutoff frequency, and a diagonal line above the cutoff frequency. There is also a "knee curve" at the boundary between the two, which smoothly transitions between the two straight line regions. If the transfer function of a first-order low-pass filter has a zero as well as a pole, the Bode plot flattens out again, at some maximum attenuation of high frequencies; such an effect is caused for example by a little bit of the input leaking around the one-pole filter; this one-pole–one-zero filter is still a first-order low-pass.

- A second-order filter attenuates high frequencies more steeply. The Bode plot for this type of filter resembles that of a first-order filter, except that it falls off more quickly. For example, a second-order Butterworth filter reduces the signal amplitude to one fourth its original level every time the frequency doubles (so power decreases by 12 dB per octave, or 40 dB per decade). Other all-pole second-order filters may roll off at different rates initially depending on their Q factor, but approach the same final rate of 12 dB per octave; as with the first-order filters, zeroes in the transfer function can change the high-frequency asymptote.
- Third- and higher-order filters are defined similarly. In general, the final rate of power roll-off for an order-n all-pole filter is $6n$ dB per octave (i.e., $20n$ dB per decade).

On any Butterworth filter, if one extends the horizontal line to the right and the diagonal line to the upper-left (the asymptotes of the function), they intersect at exactly the *cutoff frequency*. The frequency response at the cutoff frequency in a first-order filter is 3 dB below the horizontal line. The various types of filters (Butterworth filter, Chebyshev filter, Bessel filter, etc.) all have different-looking *knee curves*. Many second-order filters have "peaking" or resonance that puts their frequency response at the cutoff frequency *above* the horizontal line.

The meanings of 'low' and 'high'—that is, the cutoff frequency—depend on the characteristics of the filter. The term "low-pass filter" merely refers to the shape of the filter's response; a high-pass filter could be built that cuts off at a lower frequency than any low-pass filter—it is their responses that set them apart. Electronic circuits can be devised for any desired frequency range, right up through microwave frequencies (above 1 GHz) and higher.

Laplace Notation

Continuous-time filters can also be described in terms of the Laplace transform of their impulse response, in a way that lets all characteristics of the filter be easily analyzed by considering the pattern of poles and zeros of the Laplace transform in the complex plane. (In discrete time, one can similarly consider the Z-transform of the impulse response.)

For example, a first-order low-pass filter can be described in Laplace notation as:

$$\frac{\text{Output}}{\text{Input}} = K\frac{1}{\tau s+1}$$

where s is the Laplace transform variable, τ is the filter time constant, and K is the gain of the filter in the passband.

Electronic Low-pass Filters

First Order

One simple low-pass filter circuit consists of a resistor in series with a load, and a capacitor in parallel with the load. The capacitor exhibits reactance, and blocks low-frequency signals, forcing them through the load instead. At higher frequencies the reactance drops, and the capacitor effectively functions as a short circuit. The combination of resistance and capacitance gives the time constant of the filter $\tau = RC$ (represented by the letter tau).

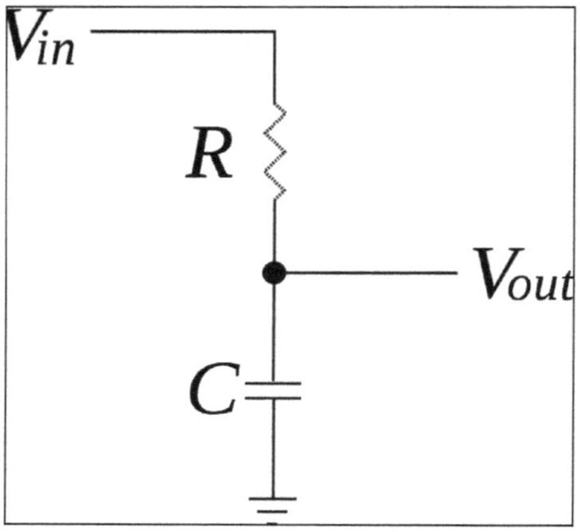

Passive, first order low-pass RC filter

The break frequency, also called the turnover frequency or cutoff frequency (in hertz), is determined by the time constant:

$$f_c = \frac{1}{2\pi\tau} = \frac{1}{2\pi RC}$$

or equivalently (in radians per second):

$$\omega_c = \frac{1}{\tau} = \frac{1}{RC}$$

This circuit may be understood by considering the time the capacitor needs to charge or discharge through the resistor:

- At low frequencies, there is plenty of time for the capacitor to charge up to practically the same voltage as the input voltage.
- At high frequencies, the capacitor only has time to charge up a small amount before the input switches direction. The output goes up and down only a small fraction of the amount the input goes up and down. At double the frequency, there's only time for it to charge up half the amount.

Another way to understand this circuit is through the concept of reactance at a particular frequency:

- Since direct current (DC) cannot flow through the capacitor, DC input must flow out the path marked V_{out} (analogous to removing the capacitor).
- Since alternating current (AC) flows very well through the capacitor, almost as well as it flows through solid wire, AC input flows out through the capacitor, effectively short circuiting to ground (analogous to replacing the capacitor with just a wire).

The capacitor is not an "on/off" object (like the block or pass fluidic explanation above). The capacitor variably acts between these two extremes. It is the Bode plot and frequency response that show this variability.

RL Filter

A resistor–inductor circuit or RL filter is an electric circuit composed of resistors and inductors driven by a voltage or current source. A first order RL circuit is composed of one resistor and one inductor and is the simplest type of RL circuit.

A first order RL circuit is one of the simplest analogue infinite impulse response electronic filters. It consists of a resistor and an inductor, either in series driven by a voltage source or in parallel driven by a current source.

Second Order

RLC Filter

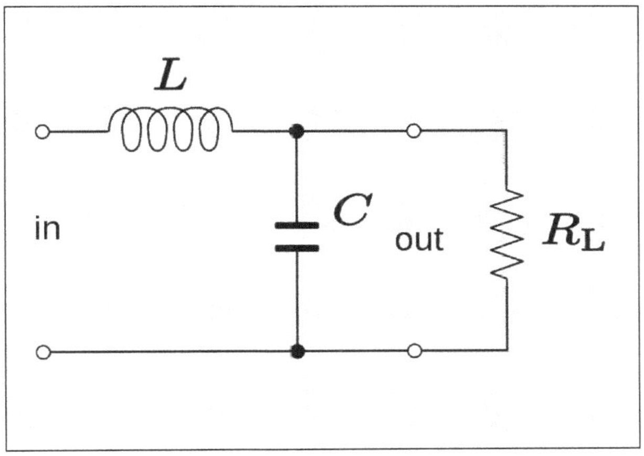

RLC circuit as a low-pass filter.

An RLC circuit (the letters R, L and C can be in other orders) is an electrical circuit consisting of a resistor, an inductor, and a capacitor, connected in series or in parallel. The RLC part of the name is due to those letters being the usual electrical symbols for resistance, inductance and capacitance respectively. The circuit forms a harmonic oscillator for current and will resonate in a similar way as an LC circuit will. The main difference that the presence of the resistor makes is that any oscillation induced in the circuit will die away over time if it is not kept going by a source. This effect of the resistor is called damping. The presence of the resistance also reduces the peak resonant frequency somewhat. Some resistance is unavoidable in real circuits, even if a resistor is not specifically included as a component. An ideal, pure LC circuit is an abstraction for the purpose of theory.

There are many applications for this circuit. They are used in many different types of oscillator circuits. Another important application is for tuning, such as in radio receivers or television sets, where they are used to select a narrow range of frequencies from the ambient radio waves. In this role the circuit is often referred to as a tuned circuit. An RLC circuit can be used as a band-pass filter, band-stop filter, low-pass filter or high-pass filter. The RLC filter is described as a *second-order* circuit, meaning that any voltage or current in the circuit can be described by a second-order differential equation in circuit analysis.

Higher Order Passive Filters

Higher order passive filters can also be constructed.

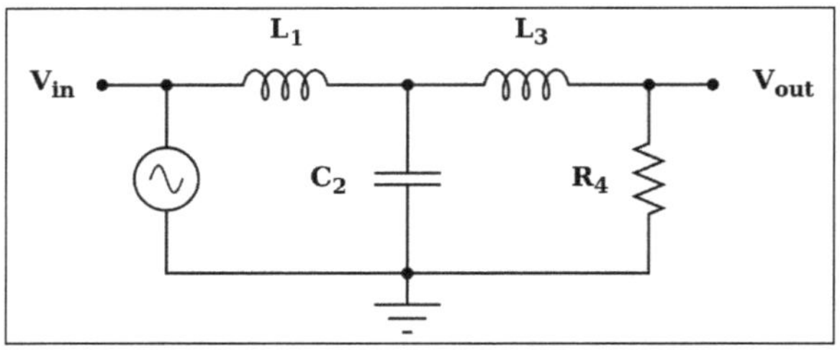

A third-order low-pass filter (Cauer topology). The filter becomes a Butterworth filter with cutoff frequency ω_c=1 when (for example) C_2=4/3 farad, R_4=1 ohm, L_1=3/2 henry and L_3=1/2 henry.

Active Electronic Realization

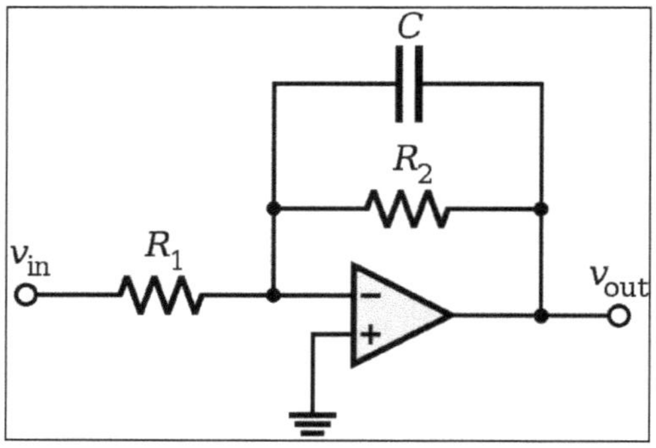

An active low-pass filter.

Another type of electrical circuit is an *active* low-pass filter.

In the operational amplifier circuit shown in the figure, the cutoff frequency (in hertz) is defined as:

$$f_c = \frac{1}{2\pi R_2 C}$$

or equivalently (in radians per second):

$$\omega_c = \frac{1}{R_2 C}$$

The gain in the passband is $-R_2/R_1$, and the stopband drops off at −6 dB per octave (that is −20 dB per decade) as it is a first-order filter.

High-pass Filter

A high-pass filter (HPF) is an electronic filter that passes signals with a frequency higher than a certain cutoff frequency and attenuates signals with frequencies lower than the cutoff frequency.

The amount of attenuation for each frequency depends on the filter design. A high-pass filter is usually modeled as a linear time-invariant system. It is sometimes called a low-cut filter or bass-cut filter. High-pass filters have many uses, such as blocking DC from circuitry sensitive to non-zero average voltages or radio frequency devices. They can also be used in conjunction with a low-pass filter to produce a bandpass filter.

In the optical domain, high-pass and low-pass have the opposite meanings, with a "high-pass" filter (more commonly "long-pass") passing only *longer* wavelengths (lower frequencies), and vice-versa for "low-pass" (more commonly "short-pass").

First-order Continuous-time Implementation

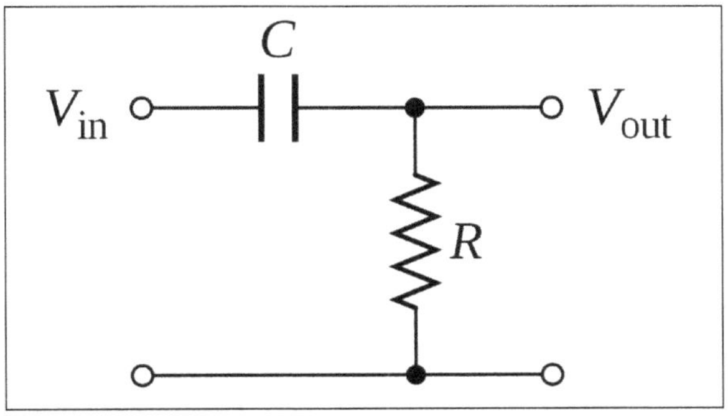

A passive, analog, first-order high-pass filter, realized by an RC circuit.

The simple first-order electronic high-pass filter shown in figure is implemented by placing an input voltage across the series combination of a capacitor and a resistor and using the voltage across the resistor as an output. The product of the resistance and capacitance ($R \times C$) is the time constant (τ); it is inversely proportional to the cutoff frequency f_c, that is,

$$f_c = \frac{1}{2\pi\tau} = \frac{1}{2\pi RC},$$

where f_c is in hertz, τ is in seconds, R is in ohms, and C is in farads.

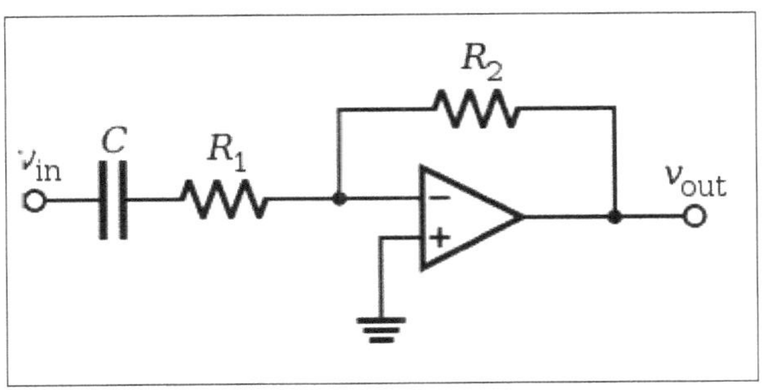

An active high-pass filter.

Figure shows an active electronic implementation of a first-order high-pass filter using an operational amplifier. In this case, the filter has a passband gain of $-R_2/R_1$ and has a cutoff frequency of:

$$f_c = \frac{1}{2\pi\tau} = \frac{1}{2\pi R_1 C},$$

Because this filter is active, it may have non-unity passband gain. That is, high-frequency signals are inverted and amplified by R_2/R_1.

Discrete-time Realization

Discrete-time high-pass filters can also be designed. a simple example comes from the conversion of the continuous-time high-pass filter above to a discrete-time realization. That is, the continuous-time behavior can be discretized.

From the circuit in figure above, according to Kirchhoff's Laws and the definition of capacitance:

$$\begin{cases} V_{\text{out}}(t) = I(t)R & \text{(V)} \\ Q_c(t) = C(V_{\text{in}}(t) - V_{\text{out}}(t)) & \text{(Q)} \\ I(t) = \dfrac{\mathrm{d}Q_c}{\mathrm{d}t} & \text{(I)} \end{cases}$$

where $Q_c(t)$ is the charge stored in the capacitor at time . Substituting Equation (Q) into Equation (I) and then Equation (I) into Equation (V) gives:

$$V_{\text{out}}(t) = \overbrace{C\left(\frac{\mathrm{d}V_{\text{in}}}{\mathrm{d}t} - \frac{\mathrm{d}V_{\text{out}}}{\mathrm{d}t}\right)}^{I(t)} R = RC\left(\frac{\mathrm{d}V_{\text{in}}}{\mathrm{d}t} - \frac{\mathrm{d}V_{\text{out}}}{\mathrm{d}t}\right)$$

This equation can be discretized. For simplicity, assume that samples of the input and output are taken at evenly spaced points in time separated by Ä_T time. Let the samples of V_{in} be represented by the sequence $(x_1, x_2, \ldots, x_n)$, and let V_{out} be represented by the sequence $(y_1, y_2, \ldots, y_n)$ which correspond to the same points in time. Making these substitutions:

$$y_i = RC\left(\frac{x_i - x_{i-1}}{\text{Ä}_T} - \frac{y_i - y_{i-1}}{\text{Ä}_T}\right)$$

And rearranging terms gives the recurrence relation

$$y_i = \overbrace{\frac{RC}{RC + \text{Ä}_T} y_{i-1}}^{\text{Decaying contribution from prior inputs}} + \overbrace{\frac{RC}{RC + \text{Ä}_T}(x_i - x_{i-1})}^{\text{Contribution from change in input}}$$

That is, this discrete-time implementation of a simple continuous-time RC high-pass filter is,

$$y_i = \alpha y_{i-1} + \alpha(x_i - x_{i-1}) \qquad \text{where} \qquad \alpha \triangleq \frac{RC}{RC + \Delta_T}$$

By definition, $0 \leq \alpha \leq 1$. The expression for parameter α yields the equivalent time constant RC in terms of the sampling period Δ_T and α:

$$RC = \Delta_T \left(\frac{\alpha}{1-\alpha} \right)..$$

Recalling that,

$$f_c = \frac{1}{2\pi RC} \text{ so } RC = \frac{1}{2\pi f_c}$$

then α and f_c are related by:

$$\alpha = \frac{1}{2\pi \Delta_T f_c + 1}$$

and

$$f_c = \frac{1-\alpha}{2\pi \alpha \Delta_T}.$$

If $\alpha = 0.5$, then the RC time constant equal to the sampling period. If $\alpha \ll 0.5$, then RC is significantly smaller than the sampling interval, and $RC \approx \alpha \Delta_T$.

Algorithmic Implementation

The filter recurrence relation provides a way to determine the output samples in terms of the input samples and the preceding output. The following pseudocode algorithm will simulate the effect of a high-pass filter on a series of digital samples, assuming equally spaced samples:

```
// Return RC high-pass filter output samples, given input samples,
// time interval dt, and time constant RC
function highpass(real[0..n] x, real dt, real RC)
 var real[0..n] y
 var real α := RC / (RC + dt)
 y := x
 for i from 1 to n
   y[i] := α * y[i-1] + α * (x[i] - x[i-1])
return y
```

The loop which calculates each of the n outputs can be refactored into the equivalent:

```
for i from 1 to n
 y[i] := α * (y[i-1] + x[i] - x[i-1])
```

However, the earlier form shows how the parameter α changes the impact of the prior output y[i-1] and current *change* in input (x[i] - x[i-1]). In particular,

- A large α implies that the output will decay very slowly but will also be strongly influenced by even small changes in input. By the relationship between parameter α and time constant RC above, a large α corresponds to a large RC and therefore a low corner frequency of the filter. Hence, this case corresponds to a high-pass filter with a very narrow stop band. Because it is excited by small changes and tends to hold its prior output values for a long time, it can pass relatively low frequencies. However, a constant input (i.e., an input with (x[i] - x[i-1])=0) will always decay to zero, as would be expected with a high-pass filter with a large .
- A small α implies that the output will decay quickly and will require large changes in the input (i.e., (x[i] - x[i-1]) is large) to cause the output to change much. By the relationship between parameter α and time constant RC above, a small α corresponds to a small RC and therefore a high corner frequency of the filter. Hence, this case corresponds to a high-pass filter with a very wide stop band. Because it requires large (i.e., fast) changes and tends to quickly forget its prior output values, it can only pass relatively high frequencies, as would be expected with a high-pass filter with a small.

Applications

Audio

High-pass filters have many applications. They are used as part of an audio crossover to direct high frequencies to a tweeter while attenuating bass signals which could interfere with, or damage, the speaker. When such a filter is built into a loudspeaker cabinet it is normally a passive filter that also includes a low-pass filter for the woofer and so often employs both a capacitor and inductor (although very simple high-pass filters for tweeters can consist of a series capacitor and nothing else). As an example, the formula above, applied to a tweeter with R=10 Ohm, will determine the capacitor value for a cut-off frequency of 5 kHz. $C = \frac{1}{2\pi fR} = \frac{1}{6.28 \times 5000 \times 10} = 3.18 \times 10^{-6}$, or approx 3.2 μF.

An alternative, which provides good quality sound without inductors (which are prone to parasitic coupling, are expensive, and may have significant internal resistance) is to employ bi-amplification with active RC filters or active digital filters with separate power amplifiers for each loudspeaker. Such low-current and low-voltage line level crossovers are called active crossovers.

Rumble filters are high-pass filters applied to the removal of unwanted sounds near to the lower end of the audible range or below. For example, noises (e.g., footsteps, or motor noises from record players and tape decks) may be removed because they are undesired or may overload the RIAA equalization circuit of the preamp.

High-pass filters are also used for AC coupling at the inputs of many audio power amplifiers, for preventing the amplification of DC currents which may harm the amplifier, rob the amplifier of headroom, and generate waste heat at the loudspeakers voice coil. One amplifier, the professional audio model DC300 made by Crown International beginning in the 1960s, did not have high-pass filtering at all, and could be used to amplify the DC signal of a common 9-volt battery at the input to supply 18 volts DC in an emergency for mixing console power. However, that model's basic design has been superseded by newer designs such as the Crown Macro-Tech series developed in the late 1980s which included 10 Hz high-pass filtering on the inputs and switchable 35 Hz high-pass filtering on the outputs. Another example is the QSC Audio PLX amplifier series which includes an internal 5 Hz high-pass filter which is applied to the inputs whenever the optional 50 and 30 Hz high-pass filters are turned off.

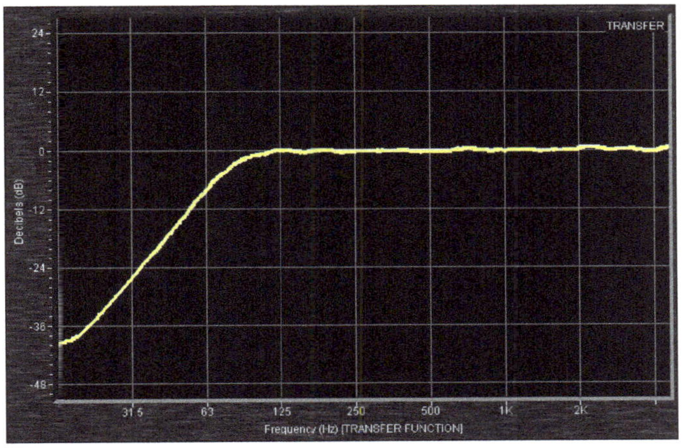

A 75 Hz "low cut" filter from an input channel of a Mackie 1402 mixing console as measured by Smaart software. This high-pass filter has a slope of 18 dB per octave.

Mixing consoles often include high-pass filtering at each channel strip. Some models have fixed-slope, fixed-frequency high-pass filters at 80 or 100 Hz that can be engaged; other models have sweepable high-pass filters, filters of fixed slope that can be set within a specified frequency range, such as from 20 to 400 Hz on the Midas Heritage 3000, or 20 to 20,000 Hz on the Yamaha M7CL digital mixing console. Veteran systems engineer and live sound mixer Bruce Main recommends that high-pass filters be engaged for most mixer input sources, except for those such as kick drum, bass guitar and piano, sources which will have useful low frequency sounds. Main writes that DI unit inputs (as opposed to microphone inputs) do not need high-pass filtering as they are not subject to modulation by low-frequency stage wash—low frequency sounds coming from the subwoofers or the public address system and wrapping around to the stage. Main indicates that high-pass filters are commonly used for directional microphones which have a proximity effect—a low-frequency boost for very close sources. This low frequency boost commonly causes problems up to 200 or 300 Hz, but Main notes that he has seen microphones that benefit from a 500 Hz high-pass filter setting on the console.

All-pass Filter

An all-pass filter is a signal processing filter that passes all frequencies equally in gain, but changes the phase relationship among various frequencies. Most types of filter reduce the amplitude (i.e. the magnitude) of the signal applied to it for some values of frequency, whereas the all-pass filter allows all frequencies through without changes in level.

Common Applications

A common application in electronic music production is in the design of an effects unit known as a "phaser", where a number of all-pass filters are connected in sequence and the output mixed with the raw signal.

It does this by varying its phase shift as a function of frequency. Generally, the filter is described by the frequency at which the phase shift crosses 90° (i.e., when the input and output signals go into quadrature – when there is a quarter wavelength of delay between them).

They are generally used to compensate for other undesired phase shifts that arise in the system, or for mixing with an unshifted version of the original to implement a notch comb filter.

They may also be used to convert a mixed phase filter into a minimum phase filter with an equivalent magnitude response or an unstable filter into a stable filter with an equivalent magnitude response.

Active Analog Implementation

Implementation using Low-pass Filter

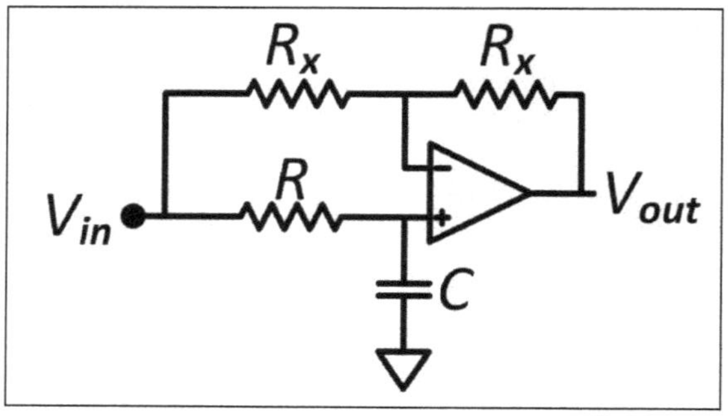

An op-amp base all-pass filter incorporating a low-pass filter.

The operational amplifier circuit shown in adjacent figure implements a single-pole active all-pass filter that features a low-pass filter at the non-inverting input of the opamp. The filter's transfer function is given by:

$$H(s) = -\frac{s - \frac{1}{RC}}{s + \frac{1}{RC}} = \frac{1 - sRC}{1 + sRC},$$

which has one pole at -1/RC and one zero at 1/RC (i.e., they are *reflections* of each other across the imaginary axis of the complex plane). The magnitude and phase of H(iω) for some angular frequency ω are

$$|H(i\omega)| = 1 \quad \text{and} \quad \angle H(i\omega) = -2\arctan(\omega RC).$$

The filter has unity-gain magnitude for all ω. The filter introduces a different delay at each frequency and reaches input-to-output *quadrature* at ω=1/RC (i.e., phase shift is 90°).

This implementation uses a low-pass filter at the non-inverting input to generate the phase shift and negative feedback.

- At high frequencies, the capacitor is a short circuit, creating an inverting amplifier (i.e., 180° phase shift) with unity gain.
- At low frequencies and DC, the capacitor is an open circuit, creating a unity-gain voltage buffer (i.e., no phase shift).
- At the corner frequency ω=1/RC of the low-pass filter (i.e., when input frequency is 1/(2πRC)), the circuit introduces a 90° shift (i.e., output is in quadrature with input; the output appears to be delayed by a quarter period from the input).

In fact, the phase shift of the all-pass filter is double the phase shift of the low-pass filter at its non-inverting input.

Interpretation as a Padé Approximation to a Pure Delay

The Laplace transform of a pure delay is given by:

$$e^{-sT},$$

where T is the delay (in seconds) and $s \in \mathbb{C}$ is complex frequency. This can be approximated using a Padé approximant, as follows:

$$e^{-sT} = \frac{e^{-sT/2}}{e^{sT/2}} \approx \frac{1 - sT/2}{1 + sT/2},$$

where the last step was achieved via a first-order Taylor series expansion of the numerator and denominator. By setting $RC = T/2$ we recover $H(s)$ from above.

Implementation using High-pass Filter

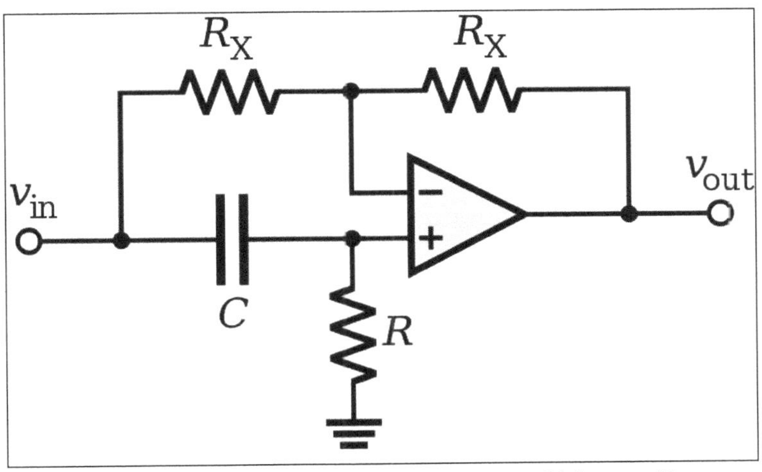

An op-amp base all-pass filter incorporating a high-pass filter.

The operational amplifier circuit shown in the adjacent figure implements a single-pole active all-pass filter that features a high-pass filter at the non-inverting input of the opamp. The filter's transfer function is given by:

$$H(s) = \frac{s - \dfrac{1}{RC}}{s + \dfrac{1}{RC}},$$

which has one pole at -1/RC and one zero at 1/RC (i.e., they are *reflections* of each other across the imaginary axis of the complex plane). The magnitude and phase of H(iω) for some angular frequency ω are,

$$|H(i\omega)| = 1 \quad \text{and} \quad \angle H(i\omega) = \pi - 2\arctan(\omega RC).$$

The filter has unity-gain magnitude for all ω. The filter introduces a different delay at each frequency and reaches input-to-output *quadrature* at ω=1/RC (i.e., phase lead is 90°).

This implementation uses a high-pass filter at the non-inverting input to generate the phase shift and negative feedback.

- At high frequencies, the capacitor is a short circuit, thereby creating a unity-gain voltage buffer (i.e., no phase lead).
- At low frequencies and DC, the capacitor is an open circuit and the circuit is an inverting amplifier (i.e., 180° phase lead) with unity gain.
- At the corner frequency ω=1/RC of the high-pass filter (i.e., when input frequency is 1/(2πRC)), the circuit introduces a 90° phase lead (i.e., output is in quadrature with input; the output appears to be advanced by a quarter period from the input).

In fact, the phase shift of the all-pass filter is double the phase shift of the high-pass filter at its non-inverting input.

Voltage Controlled Implementation

The resistor can be replaced with a FET in its *ohmic mode* to implement a voltage-controlled phase shifter; the voltage on the gate adjusts the phase shift. In electronic music, a phaser typically consists of two, four or six of these phase-shifting sections connected in tandem and summed with the original. A low-frequency oscillator (LFO) ramps the control voltage to produce the characteristic swooshing sound.

Passive Analog Implementation

The benefit to implementing all-pass filters with active components like operational amplifiers is that they do not require inductors, which are bulky and costly in integrated circuit designs. In other applications where inductors are readily available, all-pass filters can be implemented entirely without active components. There are a number of circuit topologies that can be used for this. The following are the most commonly used circuits.

Lattice Filter

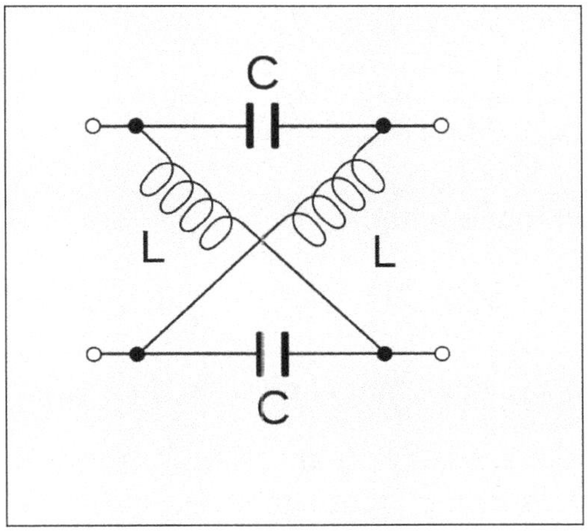

An all-pass filter using lattice topology.

The lattice phase equaliser, or filter, is a filter composed of lattice, or X-sections. With single element branches it can produce a phase shift up to 180°, and with resonant branches it can produce phase shifts up to 360°. The filter is an example of a constant-resistance network (i.e., its image impedance is constant over all frequencies).

T-section Filter

The phase equaliser based on T topology is the unbalanced equivalent of the lattice filter and has the same phase response. While the circuit diagram may look like a low pass filter it is different in that the two inductor branches are mutually coupled. This results in transformer action between the two inductors and an all-pass response even at high frequency.

Bridged T-section Filter

The bridged T topology is used for delay equalisation, particularly the differential delay between two landlines being used for stereophonic sound broadcasts. This application requires that the filter has a linear phase response with frequency (i.e., constant group delay) over a wide bandwidth and is the reason for choosing this topology.

Digital Implementation

A Z-transform implementation of an all-pass filter with a complex pole at z_0 is,

$$H(z) = \frac{z^{-1} - \overline{z_0}}{1 - z_0 z^{-1}}$$

which has a zero at $1/\overline{z_0}$, where $\overline{z}$ denotes the complex conjugate. The pole and zero sit at the same angle but have reciprocal magnitudes (i.e., they are *reflections* of each other across the boundary of the complex unit circle). The placement of this pole-zero pair for a given z_0 can be rotated in the complex plane by any angle and retain its all-pass magnitude characteristic. Complex pole-zero pairs in all-pass filters help control the frequency where phase shifts occur.

To create an all-pass implementation with real coefficients, the complex all-pass filter can be cascaded with an all-pass that substitutes $\overline{z_0}$ for z_0, leading to the Z-transform implementation

$$H(z) = \frac{z^{-1} - \overline{z_0}}{1 - z_0 z^{-1}} \times \frac{z^{-1} - z_0}{1 - \overline{z_0} z^{-1}} = \frac{z^{-2} - 2\Re(z_0) z^{-1} + |z_0|^2}{1 - 2\Re(z_0) z^{-1} + |z_0|^2 z^{-2}},$$

which is equivalent to the difference equation

$$y[k] - 2\Re(z_0) y[k-1] + |z_0|^2 y[k-2] = x[k-2] - 2\Re(z_0) x[k-1] + |z_0|^2 x[k],$$

where $y[k]$ is the output and $x[k]$ is the input at discrete time step .

Filters such as the above can be cascaded with unstable or mixed-phase filters to create a stable or minimum-phase filter without changing the magnitude response of the system. For example, by proper choice of z_0, a pole of an unstable system that is outside of the unit circle can be canceled and reflected inside the unit circle.

Anti-aliasing Filter

An anti-aliasing filter (AAF) is a filter used before a signal sampler to restrict the bandwidth of a signal to approximately or completely satisfy the Nyquist–Shannon sampling theorem over the band of interest. Since the theorem states that unambiguous reconstruction of the signal from its samples is possible when the power of frequencies above the Nyquist frequency is zero, a real anti-aliasing filter trades off between bandwidth and aliasing. A realizable anti-aliasing filter will typically either permit some aliasing to occur or else attenuate some in-band frequencies close to the Nyquist limit. For this reason, many practical systems sample higher than would be theoretically required by a perfect AAF in order to ensure that all frequencies of interest can be reconstructed, a practice called oversampling.

Optical Applications

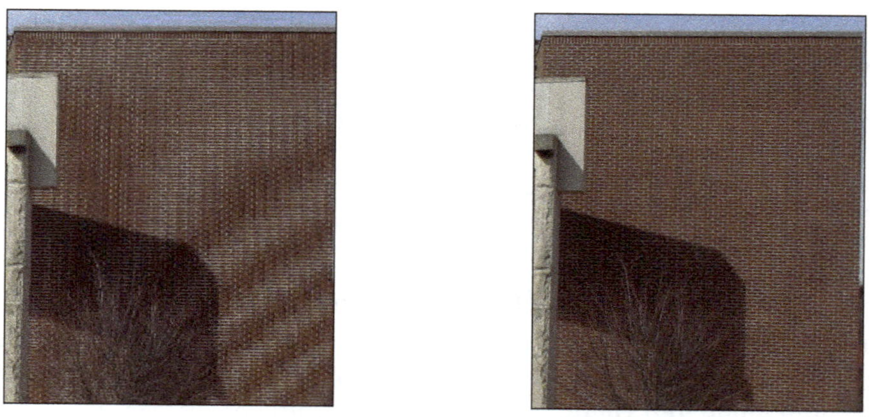

Simulated photographs of a brick wall without (left) and with (right) an optical low-pass filter.

In the case of optical image sampling, as by image sensors in digital cameras, the anti-aliasing filter is also known as an optical low-pass filter (OLPF), blur filter, or AA filter. The mathematics of sampling in two spatial dimensions is similar to the mathematics of time-domain sampling, but the filter implementation technologies are different. The typical implementation in digital cameras is two layers of birefringent material such as lithium niobate, which spreads each optical point into a cluster of four points.

The choice of spot separation for such a filter involves a tradeoff among sharpness, aliasing, and fill factor (the ratio of the active refracting area of a microlens array to the total contiguous area occupied by the array). In a monochrome or three-CCD or Foveon X3 camera, the microlens array alone, if near 100% effective, can provide a significant anti-aliasing effect, while in color filter array (CFA, e.g. Bayer filter) cameras, an additional filter is generally needed to reduce aliasing to an acceptable level.

The Pentax K-3 from Ricoh introduced a unique sensor-based anti-aliasing filter. The filter works by micro vibrating the sensor element. The user can turn the vibration on or off, selecting anti-aliasing or no anti-aliasing.

Audio Applications

Anti-aliasing filters are commonly used at the input of digital signal processing system's analog to digital converter; similar filters are used as reconstruction filters at the output of such systems, for example in music players. In the latter case, the filter prevents imaging, the reverse process of aliasing where in-band frequencies are mirrored out of band.

Oversampling

A technique known as oversampling is commonly used in audio ADCs. The idea is to use a higher intermediate digital sample rate, so that a nearly ideal digital filter can sharply cut off aliasing near the original low Nyquist frequency and give better phase response, while a much simpler analog filter can stop frequencies above the new higher Nyquist frequency. Because analog filters have relatively high cost and limited performance, relaxing the demands on the analog filter can greatly reduce both aliasing and cost. Furthermore, because some noise is averaged out, the higher sampling rate can moderately improve SNR.

Alternatively, a signal may be intentionally oversampled without an intermediate frequency to reduce the requirements on the anti-alias filter. For example, CD audio typically extends up to 20 kHz, but is sampled with a 22.05 kHz Nyquist rate. By oversampling by 2.05 kHz, both aliasing and attenuation of higher audio frequencies can be prevented even with less than ideal filters.

Bandpass Signals

Often, an anti-aliasing filter is a low-pass filter; this is not a requirement, however. Generalizations of the Nyquist–Shannon sampling theorem allow sampling of other band-limited passband signals instead of baseband signals.

For signals that are bandwidth limited, but not centered at zero, a band-pass filter can be used as an anti-aliasing filter. For example, this could be done with a single-sideband modulated or frequency modulated signal. If one desired to sample an FM radio broadcast centered at 87.9 MHz and bandlimited to a 200 kHz band, then an appropriate anti-alias filter would be centered on 87.9 MHz with 200 kHz bandwidth (or pass-band of 87.8 MHz to 88.0 MHz), and the sampling rate would be no less than 400 kHz, but should also satisfy other constraints to prevent aliasing.

Signal Overload

It is very important to avoid input signal overload when using an anti-aliasing filter. If the signal is strong enough, it can cause clipping at the analog-to-digital converter, even after filtering. When distortion due to clipping occurs after the anti-aliasing filter, it can create components outside the passband of the anti-aliasing filter; these components can then alias, causing the reproduction of other non-harmonically related frequencies.

Band-pass Filter

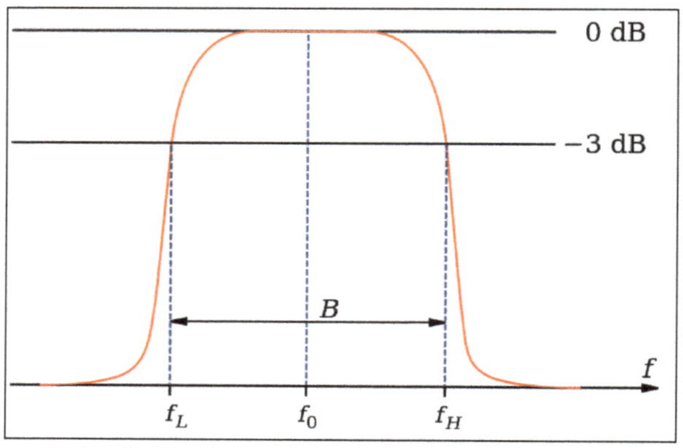

Bandwidth measured at half-power points (gain -3 dB, √2/2, or about 0.707 relative to peak) on a diagram showing magnitude transfer function versus frequency for a band-pass filter.

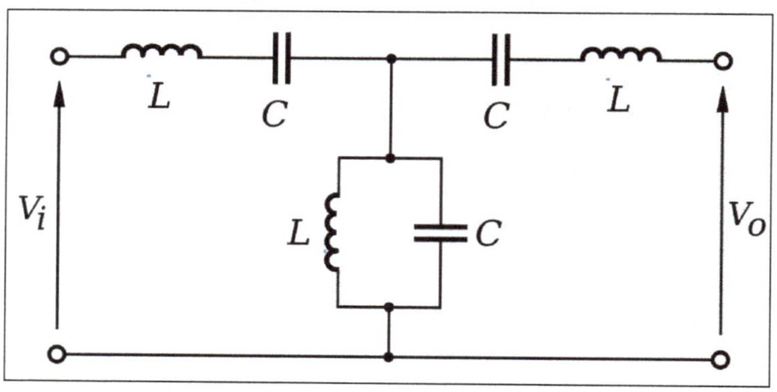

A medium-complexity example of a band-pass filter.

A band-pass filter, also bandpass filter or BPF, is a device that passes frequencies within a certain range and rejects (attenuates) frequencies outside that range.

An example of an analogue electronic band-pass filter is an RLC circuit (a resistor–inductor–capacitor circuit). These filters can also be created by combining a low-pass filter with a high-pass filter.

Bandpass is an adjective that describes a type of filter or filtering process; it is to be distinguished from passband, which refers to the actual portion of affected spectrum. Hence, one might say "A dual bandpass filter has two passbands." A *bandpass signal* is a signal containing a band of frequencies not adjacent to zero frequency, such as a signal that comes out of a bandpass filter.

An ideal bandpass filter would have a completely flat passband (e.g. with no gain/attenuation throughout) and would completely attenuate all frequencies outside the passband. Additionally, the transition out of the passband would have brickwall characteristics.

In practice, no bandpass filter is ideal. The filter does not attenuate all frequencies outside the desired frequency range completely; in particular, there is a region just outside the intended passband where frequencies are attenuated, but not rejected. This is known as the filter roll-off, and it is usually expressed in dB of attenuation per octave or decade of frequency. Generally, the design of a filter seeks to make the roll-off as narrow as possible, thus allowing the filter to perform as close as possible to its intended design. Often, this is achieved at the expense of pass-band or stop-band *ripple*.

The bandwidth of the filter is simply the difference between the upper and lower cutoff frequencies. The shape factor is the ratio of bandwidths measured using two different attenuation values to determine the cutoff frequency, e.g., a shape factor of 2:1 at 30/3 dB means the bandwidth measured between frequencies at 30 dB attenuation is twice that measured between frequencies at 3 dB attenuation.

Optical band-pass filters are common in photography and theatre lighting work. These filters take the form of a transparent coloured film or sheet.

Q Factor

A band-pass filter can be characterized by its Q factor. The Q-factor is the reciprocal of the fractional bandwidth. A high-Q filter will have a narrow passband and a low-Q filter will have a wide passband. These are respectively referred to as *narrow-band* and *wide-band* filters.

Applications

Bandpass filters are widely used in wireless transmitters and receivers. The main function of such a filter in a transmitter is to limit the bandwidth of the output signal to the band allocated for the transmission. This prevents the transmitter from interfering with other stations. In a receiver, a bandpass filter allows signals within a selected range of frequencies to be heard or decoded, while preventing signals at unwanted frequencies from getting through. A bandpass filter also optimizes the signal-to-noise ratio and sensitivity of a receiver.

In both transmitting and receiving applications, well-designed bandpass filters, having the optimum bandwidth for the mode and speed of communication being used, maximize the number of signal transmitters that can exist in a system, while minimizing the interference or competition among signals.

Outside of electronics and signal processing, one example of the use of band-pass filters is in the atmospheric sciences. It is common to band-pass filter recent meteorological data with a period range of, for example, 3 to 10 days, so that only cyclones remain as fluctuations in the data fields.

Loudspeaker Enclosures

Compound or Band-pass

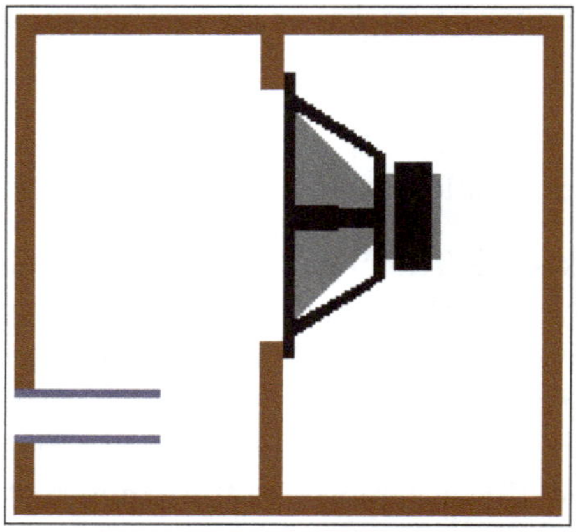

Compound or 4th order band-pass enclosure.

A 4th order electrical bandpass filter can be simulated by a vented box in which the contribution from the rear face of the driver cone is trapped in a sealed box, and the radiation from the front surface of the cone is into a ported chamber. This modifies the resonance of the driver. In its simplest form a compound enclosure has two chambers. The dividing wall between the chambers holds the driver; typically only one chamber is ported.

If the enclosure on each side of the woofer has a port in it then the enclosure yields a 6th order band-pass response. These are considerably harder to design and tend to be very sensitive to driver characteristics. As in other reflex enclosures, the ports may generally be replaced by passive radiators if desired.

An eighth order bandpass box is another variation which also has a narrow frequency range. They are often used to achieve sound pressure levels in which case a bass tone of a specific frequency would be used versus anything musical. They are complicated to build and must be done quite precisely in order to perform nearly as intended.

Other Fields

In neuroscience, visual cortical simple cells were first shown by David Hubel and Torsten Wiesel to have response properties that resemble Gabor filters, which are band-pass.

In astronomy, band-pass filters are used to allow only a single portion of the light spectrum into an instrument. Band-pass filters can help with finding where stars lie on the main sequence, identifying redshifts, and many other applications.

Band-stop Filter

The Band Stop Filter, (BSF) is another type of frequency selective circuit that functions in exactly the opposite way to the Band Pass Filter we looked at before. The band stop filter, also known as a band reject filter, passes all frequencies with the exception of those within a specified stop band which are greatly attenuated.

If this stop band is very narrow and highly attenuated over a few hertz, then the band stop filter is more commonly referred to as a notch filter, as its frequency response shows that of a deep notch with high selectivity (a steep-side curve) rather than a flattened wider band.

Also, just like the band pass filter, the band stop (band reject or notch) filter is a second-order (two-pole) filter having two cut-off frequencies, commonly known as the -3dB or half-power points producing a wide stop band bandwidth between these two -3dB points.

Then the function of a band stop filter is too pass all those frequencies from zero (DC) up to its first (lower) cut-off frequency point f_L, and pass all those frequencies above its second (upper) cut-off frequency f_H, but block or reject all those frequencies in-between. Then the filters bandwidth, BW is defined as: $(f_H - f_L)$.

So for a wide-band band stop filter, the filters actual stop band lies between its lower and upper -3dB points as it attenuates, or rejects any frequency between these two cut-off frequencies. The frequency response curve of an ideal band stop filter is therefore given as:

Band Stop Filter Response

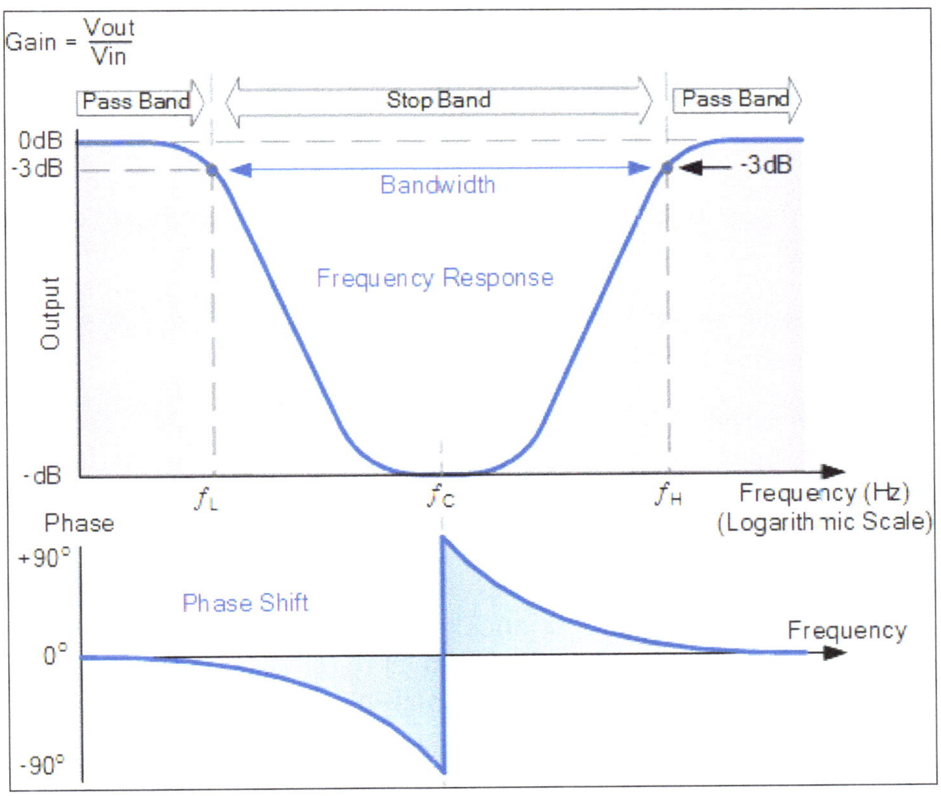

We can see from the amplitude and phase curves above for the band pass circuit, that the quantities f_L, f_H and f_C are the same as those used to describe the behaviour of the band-pass filter. This is because the band stop filter is simply an inverted or complimented form of the standard band-pass filter. In fact the definitions used for bandwidth, pass band, stop band and center frequency are the same as before, and we can use the same formulas to calculate bandwidth, BW, center frequency, f_C, and quality factor, Q.

The ideal band stop filter would have infinite attenuation in its stop band and zero attenuation in either pass band. The transition between the two pass bands and the stop band would be vertical (brick wall). There are several ways we can design a "Band Stop Filter", and they all accomplish the same purpose.

Generally band-pass filters are constructed by combining a low pass filter (LPF) in series with a high pass filter (HPF). Band stop filters are created by combining together the low pass and high pass filter sections in a "parallel" type configuration as shown.

Typical Band Stop Filter Configuration

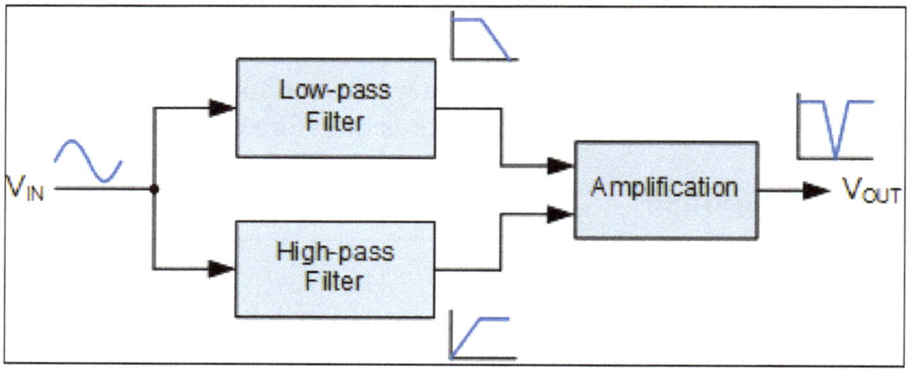

The summing of the high pass and low pass filters means that their frequency responses do not overlap, unlike the band-pass filter. This is due to the fact that their start and ending frequencies are at different frequency points. For example, suppose we have a first-order low-pass filter with a cut-off frequency, f of 200Hz connected in parallel with a first-order high-pass filter with a cut-off frequency, f_H of 800Hz. As the two filters are effectively connected in parallel, the input signal is applied to both filters simultaneously as shown above.

All of the input frequencies below 200Hz would be passed unattenuated to the output by the low-pass filter. Likewise, all input frequencies above 800Hz would be passed unattenuated to the output by the high-pass filter. However, and input signal frequencies in-between these two frequency cut-off points of 200Hz and 800Hz, that is f_L to f_H would be rejected by either filter forming a notch in the filters output response.

In other words a signal with a frequency of 200Hz or less and 800Hz and above would pass unaffected but a signal frequency of say 500Hz would be rejected as it is too high to be passed by the low-pass filter and too low to be passed by the high-pass filter. We can show the effect of this frequency characteristic below.

Band Stop Filter Characteristics

The transformation of this filter characteristic can be easily implemented using a single low pass and high pass filter circuits isolated from each other by non-inverting voltage follower, (Av = 1).

The output from these two filter circuits is then summed using a third operational amplifier connected as a voltage summer (adder) as shown.

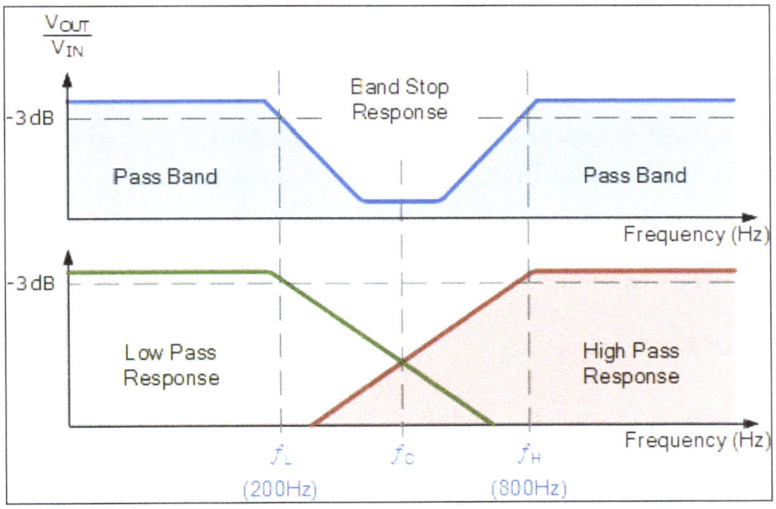

Band Stop Filter Circuit

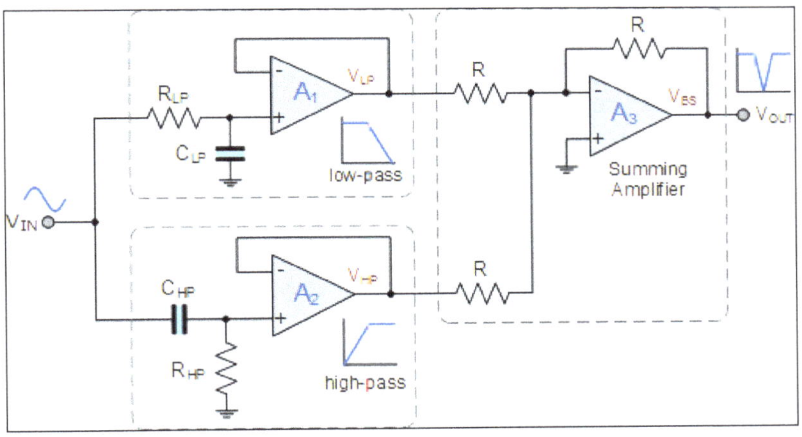

The use of operational amplifiers within the band stop filter design also allows us to introduce voltage gain into the basic filter circuit. The two non-inverting voltage followers can easily be converted into a basic non-inverting amplifier with a gain of $Av = 1 + Rf / Rin$ by the addition of input and feedback resistors.

Also if we require a band stop filter to have its -3dB cut-off points at say, 1kHz and 10kHz and a stop band gain of -10dB in between, we can easily design a low-pass filter and a high-pass filter with these requirements and simply cascade them together to form our wide-band band-pass filter design.

Now we understand the principle behind a Band Stop Filter, let us design one using the previous cut-off frequency values.

Band Stop Filter Example

Design a basic wide-band, RC band stop filter with a lower cut-off frequency of 200Hz and a higher cut-off frequency of 800Hz. Find the geometric center frequency, -3dB bandwidth and Q of the circuit.

$$f = \frac{1}{2\pi RC} HZ$$

The upper and lower cut-off frequency points for a band stop filter can be found using the same formula as that for both the low and high pass filters as shown.

Assuming a capacitor, C value for both filter sections of 0.1uF, the values of the two frequency determining resistors, R_L and R_H are calculated as follows.

Low Pass Filter Section

$$f_L = \frac{1}{2\pi R_L C} = 200\text{Hz} \quad \text{and} \quad C = 0.1\text{uF}$$

$$\therefore R_L = \frac{1}{2\pi \times 200 \times 0.1 \times 10^{-6}} = 7958\Omega \text{ or } 8k\Omega$$

High Pass Filter Section

$$f_H = \frac{1}{2\pi R_H C} = 800\text{Hz} \quad \text{and} \quad C = 0.1\text{uF}$$

$$\therefore R_H = \frac{1}{2\pi \times 800 \times 0.1 \times 10^{-6}} = 1990\Omega \text{ or } 2k\Omega$$

From this we can calculate the geometric center frequency, f_C as:

$$f_C = \sqrt{f_L \times f_H} = \sqrt{200 \times 80} = 400\text{Hz}$$

$$f_{BW} = f_H - f_L = 800 - 200 = 600\text{Hz}$$

$$Q = \frac{f_C}{f_{BW}} = \frac{400}{600} = 0.67 \text{ or } -3.5\text{dB}$$

Now that we know the component values for the two filter stages, we can combine them into a single voltage adder circuit to complete our filter design. The magnitude and polarity of the adders output will be at any given time, the algebraic sum of its two inputs.

If we make the op-amps feedback resistor and its two input resistors the same values, say 10kΩ, then the inverting summing circuit will provide a mathematically correct sum of the two input signals with zero voltage gain.

Band Stop Filter Design

simple band stop filters can be made using first or second order low and high pass filters along with a non-inverting summing op-amp circuit to reject a wide band of frequencies. But we can also design and construct band stop filters to produce a much narrower frequency response to eliminate specific frequencies by increasing the selectivity of the filter. This type of filter design is called a "Notch Filter".

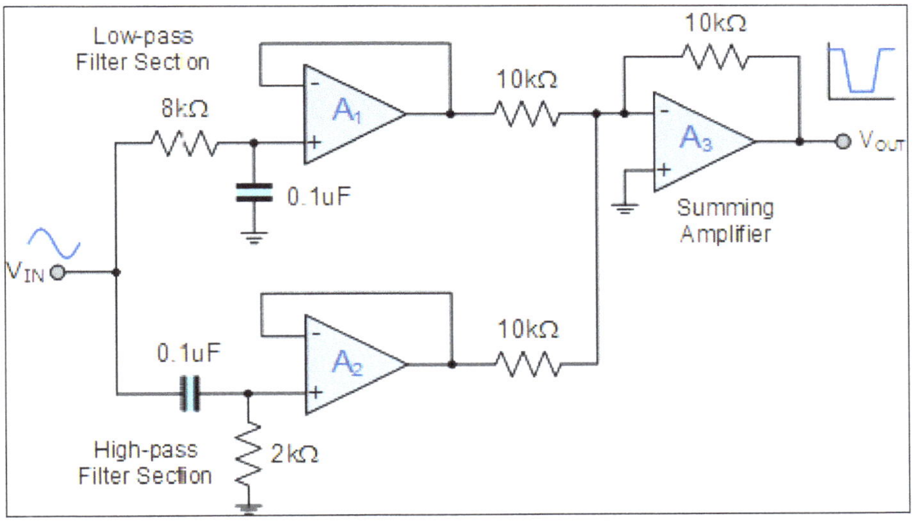

Notch Filters

Notch filters are a highly selective, high-Q, form of the band stop filter which can be used to reject a single or very small band of frequencies rather than a whole bandwidth of different frequencies. For example, it may be necessary to reject or attenuate a specific frequency generating electrical noise (such as mains hum) which has been induced into a circuit from inductive loads such as motors or ballast lighting, or the removal of harmonics, etc.

But as well as filtering, variable notch filters are also used by musicians in sound equipment such as graphic equalizers, synthesizers and electronic crossovers to deal with narrow peaks in the acoustic response of the music. Then we can see that notch filters are widely used in much the same way as low-pass and high-pass filters.

Notch filters by design have a very narrow and very deep stop band around their center frequency with the width of the notch being described by its selectivity Q in exactly the same way as resonance frequency peaks in RLC circuits.

The most common notch filter design is the twin-T notch filter network. In its basic form, the twin-T, also called a parallel-tee, configuration consists of two RC branches in the form of two tee sections, that use three resistors and three capacitors with opposite and opposing R and C elements in the tee part of its design as shown, creating a deeper notch.

Basic Twin-t Notch Filter Design

The upper T-pad configuration of resistors 2R and capacitor 2C form the low-pass filter section of the design, while the lower T-pad configuration of capacitors C and resistor R form the high-pass filter section. The frequency at which this basic twin-T notch filter design offers maximum attenuation is called the "notch frequency", f_N.

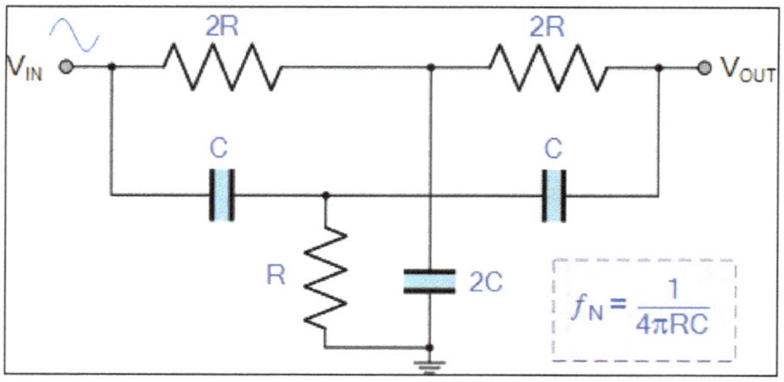

Twin-T Notch Filter Equation

$$f_N = \frac{1}{4\pi RC}$$

Being a passive RC network, one of the disadvantages of this basic twin-T notch filter design is that the maximum value of the output (Vout) below the notch frequency is generally less than the maximum value of output above the notch frequency due in part to the two series resistances (2R) in the low-pass filter section having greater losses than the reactances of the two series capacitors (C) in the high-pass section.

As well as uneven gains either side of the notch frequency, another disadvantage of this basic design is that it has a fixed Q value of 0.25, in the order of -12dB. This is because at the notch frequency, the reactances of the two series capacitors equals the resistances of the two series resistors, resulting in the currents flowing in each branch being out-of-phase by 1800.

We can improve on this by making the notch filter more selective with the application of positive feedback connected to the center of the two reference legs. Instead of connecting the junction of R and 2C to ground, (0v) but instead connect it to the central pin of a voltage divider network powered by the output signal, the amount of the signal feedback, set by the voltage divider ratio, determines the value of Q, which in turn, determines to some extent, the depth of the notch.

Single Op-amp Twin-T Notch Filter

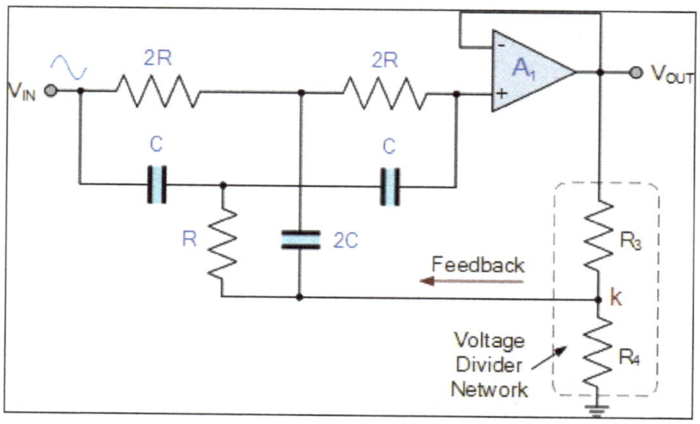

Here the output from the twin-T notch filter section is isolated from the voltage divider by a single non-inverting op-amp buffer. The output from the voltage divider is fed back to "ground" point of R and 2C. The amount of signal feedback, known as the feedback fraction k, is set by the resistor ratio and is given as:

$$k = \frac{R_4}{R_3 + R_4} = 1 - \frac{1}{4Q}$$

The value of Q is determined by the R3 and R4 resistor ratio, but if we wanted to make Q fully adjustable, we could replace these two feedback resistors with a single potentiometer and feed it into another op-amp buffer for increased negative gain. Also, to obtain the maximum notch depth at the given frequency, resistors R3 and R4 could be eliminated and the junction of R and 2C connected directly to the output.

Band Stop Filter Example

Design a two op-amp narrow-band, RC notch filter with a center notch frequency, *f*N of 1kHz and a -3dB bandwidth of 100 Hz. Use 0.1uF capacitors in your design and calculate the expected notch depth in decibels.

Data given: f_N = 1000Hz, BW = 100Hz and C = 0.1uF.

1. Calculate value of R for the given capacitance of 0.1uF,

$$R = \frac{1}{4\pi f_N C} = \frac{1}{4\pi \times 1000 \times 0.1 \times 10^{-6}}$$

$$\therefore R = 795\Omega \text{ or } 800\Omega$$

2. Calculate value of Q,

$$Q = \frac{f_N}{BW} = \frac{1000}{100} = 10$$

3. Calculate value of feedback fraction k,

$$k = 1 - \frac{1}{4Q} = 1 - \frac{1}{4 \times 10} = 0.975$$

4. Calculate the values of resistors R3 and R4,

$$k = 0.975 = \frac{R_4}{R_3 + R_4}$$

Assume $R_4 = 10k\Omega$, then R_3 equals:

$$R_3 = R_4 = 0.975R_4 = 10000 - 0.975 \times 10000$$

$$\therefore R_3 = 250\Omega$$

5. Calculate expected notch depth in decibels, dB

$$\frac{1}{Q}=\frac{1}{10}=0.1$$

$$f_{N(dB)}=20\log(0.1)=-20dB$$

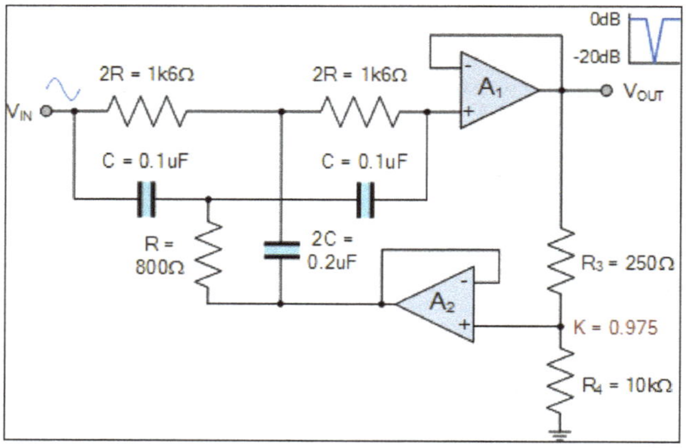

Notch filter design.

Chebyshev Filter

Chebyshev filters are analog or digital filters having a steeper roll-off than Butterworth filters, and have passband ripple (type I) or stopband ripple (type II). Chebyshev filters have the property that they minimize the error between the idealized and the actual filter characteristic over the range of the filter, but with ripples in the passband. This type of filter is named after Pafnuty Chebyshev because its mathematical characteristics are derived from Chebyshev polynomials. The type I Chebyshev filters are called usually as just "Chebyshev filters", the type II ones are usually called "inverse Chebyshev filters".

Because of the passband ripple inherent in Chebyshev filters, the ones that have a smoother response in the passband but a more irregular response in the stopband are preferred for some applications.

Type I Chebyshev Filters (Chebyshev Filters)

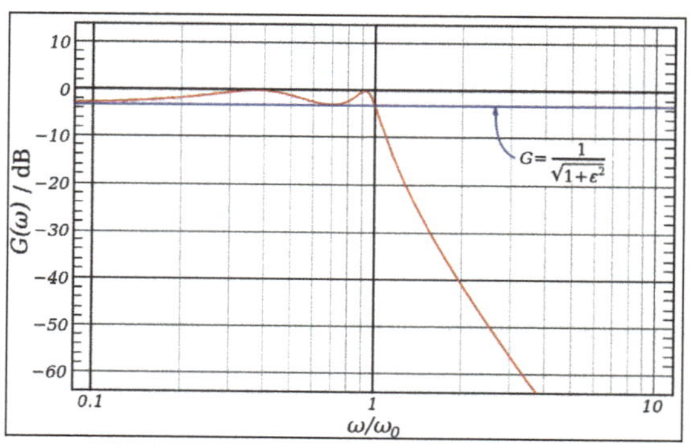

The frequency response of a fourth-order type I Chebyshev low-pass filter with $\varepsilon = 1$.

Type I Chebyshev filters are the most common types of Chebyshev filters. The gain (or amplitude) response, $G_n(\omega)$, as a function of angular frequency ω of the nth-order low-pass filter is equal to the absolute value of the transfer function $H_n(s)$ evaluated at $s = j\omega$:

$$G_n(\omega) = |H_n(j\omega)| = \frac{1}{\sqrt{1+\varepsilon^2 T_n^2\left(\frac{\omega}{\omega_0}\right)}}$$

where ε is the ripple factor, ω_0 is the cutoff frequency and T_n is a Chebyshev polynomial of the n th order.

The passband exhibits equiripple behavior, with the ripple determined by the ripple factor ε. In the passband, the Chebyshev polynomial alternates between -1 and 1 so the filter gain alternate between maxima at G = 1 and minima at $G = 1/\sqrt{1+\varepsilon^2}$.

The ripple factor ε is thus related to the passband ripple in decibels by:

$$\varepsilon = \sqrt{10^{\delta/10} - 1}.$$

At the cutoff frequency ω_0 the gain again has the value $1/\sqrt{1+\varepsilon^2}$ but continues to drop into the stopband as the frequency increases. This behavior is shown in the diagram on the right. The common practice of defining the cutoff frequency at −3 dB is usually not applied to Chebyshev filters; instead the cutoff is taken as the point at which the gain falls to the value of the ripple for the final time.

The 3 dB frequency ω_H is related to ω_0 by:

$$\omega_H = \omega_0 \cosh\left(\frac{1}{n}\cosh^{-1}\frac{1}{\varepsilon}\right).$$

The order of a Chebyshev filter is equal to the number of reactive components (for example, inductors) needed to realize the filter using analog electronics.

An even steeper roll-off can be obtained if ripple is allowed in the stopband, by allowing zeroes on the $j\omega$-axis in the complex plane. However, this results in less suppression in the stopband. The result is called an elliptic filter, also known as Cauer filter.

Poles and Zeroes

Log of the absolute value of the gain of an 8th order Chebyshev type I filter in complex frequency space (s = σ + jω) with ε = 0.1 and $\omega_0 = 1$. The white spots are poles and are arranged on an ellipse with a semi-axis of 0.3836 in σ and 1.071 in ω. The transfer function poles are those poles in the left half plane. Black corresponds to a gain of 0.05 or less, white corresponds to a gain of 20 or more.

For simplicity, it is assumed that the cutoff frequency is equal to unity. The poles (ω_{pm}) of the gain function of the Chebyshev filter are the zeroes of the denominator of the gain function.

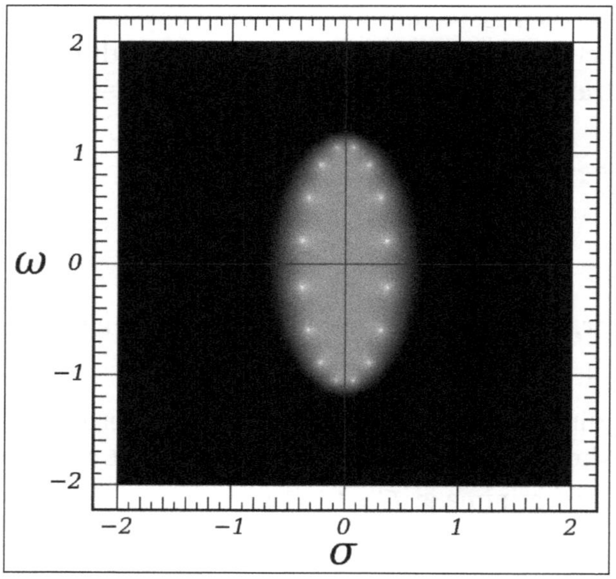

Using the complex frequency s, these occur when:

$$1+\varepsilon^2 T_n^2(-js)=0.$$

Defining $-js=\cos(\theta)$ and using the trigonometric definition of the Chebyshev polynomials yields:

$$1+\varepsilon^2 T_n^2(\cos(\theta))=1+\varepsilon^2\cos^2(n\theta)=0$$

Solving for θ,

$$\theta=\frac{1}{n}\arccos\left(\frac{\pm j}{\varepsilon}\right)+\frac{m\pi}{n}$$

where the multiple values of the arc cosine function are made explicit using the integer index m. The poles of the Chebyshev gain function are then:

$$s_{pm}=j\cos(\theta)$$

$$=j\cos\left(\frac{1}{n}\arccos\left(\frac{\pm j}{\varepsilon}\right)+\frac{m\pi}{n}\right).$$

Using the properties of the trigonometric and hyperbolic functions, this may be written in explicitly complex form:

$$s_{pm}^{\pm}=\pm\sinh\left(\frac{1}{n}\operatorname{arsinh}\left(\frac{1}{\varepsilon}\right)\right)\sin(\theta_m)$$

$$+j\cosh\left(\frac{1}{n}\operatorname{arsinh}\left(\frac{1}{\varepsilon}\right)\right)\cos(\theta_m)$$

where m = 1, 2,..., n and

$$\theta_m = \frac{\pi}{2}\frac{2m-1}{n}.$$

This may be viewed as an equation parametric in θ_n and it demonstrates that the poles lie on an ellipse in s-space centered at s = 0 with a real semi-axis of length $\sinh(\mathrm{arsinh}(1/\varepsilon)/n)$ and an imaginary semi-axis of length of $\cosh(\mathrm{arsinh}(1/\varepsilon)/n)$.

The Transfer Function

The above expression yields the poles of the gain G. For each complex pole, there is another which is the complex conjugate, and for each conjugate pair there are two more that are the negatives of the pair. The transfer function must be stable, so that its poles are those of the gain that have negative real parts and therefore lie in the left half plane of complex frequency space. The transfer function is then given by

$$H(s) = \frac{1}{2^{n-1}\varepsilon}\prod_{m=1}^{n}\frac{1}{(s - s_{pm}^{-})}$$

where s_{pm}^{-} are only those poles of the gain with a negative sign in front of the real term in the above equation for the poles.

The Group Delay

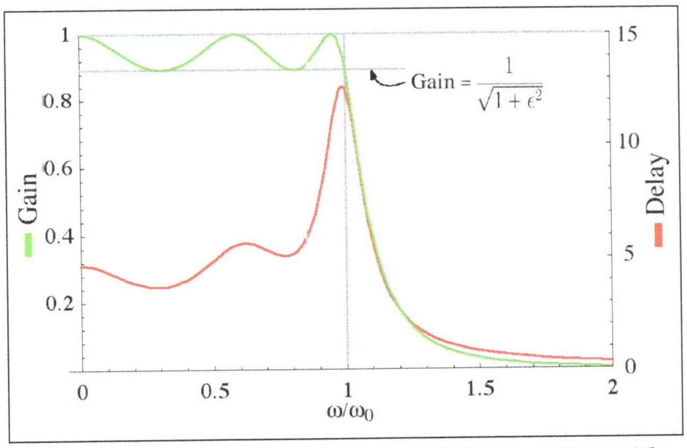

Gain and group delay of a fifth-order type I Chebyshev filter with ε = 0.5.

The group delay is defined as the derivative of the phase with respect to angular frequency and is a measure of the distortion in the signal introduced by phase differences for different frequencies.

$$\tau_g = -\frac{d}{d\omega}\arg(H(j\omega))$$

The gain and the group delay for a fifth-order type I Chebyshev filter with ε=0.5 are plotted in the graph on the left. It can be seen that there are ripples in the gain and the group delay in the pass-band but not in the stopband.

Type II Chebyshev Filters (Inverse Chebyshev Filters)

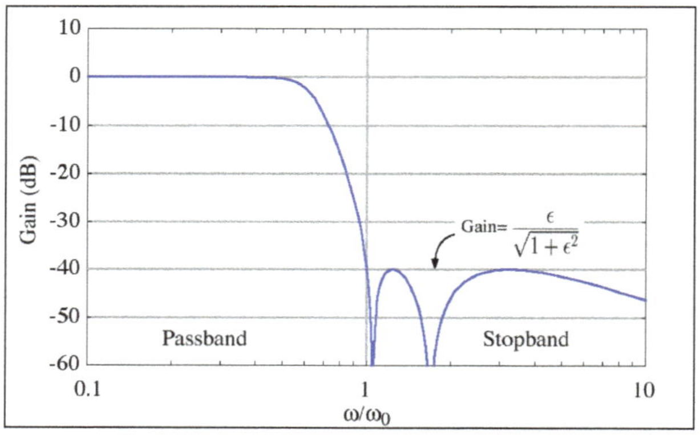

The frequency response of a fifth-order type II Chebyshev low-pass filter with $\varepsilon = 0.01$.

$$G_n(\omega, \omega_0) = \frac{1}{\sqrt{1 + \frac{1}{\varepsilon^2 T_n^2(\omega_0/\omega)}}}.$$

In the stopband, the Chebyshev polynomial oscillates between -1 and 1 so that the gain will oscillate between zero and

$$\frac{1}{\sqrt{1 + \frac{1}{\varepsilon^2}}}$$

and the smallest frequency at which this maximum is attained is the cutoff frequency ω_o. The parameter ε is thus related to the stopband attenuation γ in decibels by:

$$\varepsilon = \frac{1}{\sqrt{10^{\gamma/10} - 1}}.$$

For a stopband attenuation of 5 dB, ε = 0.6801; for an attenuation of 10 dB, ε = 0.3333. The frequency $f_o = \omega_o/2\pi$ is the cutoff frequency. The 3 dB frequency f_H is related to f_o by:

$$f_H = \frac{f_0}{\cosh\left(\frac{1}{n}\cosh^{-1}\frac{1}{\varepsilon}\right)}.$$

Poles and Zeroes

Log of the absolute value of the gain of an 8th order Chebyshev type II filter in complex frequency space (s=σ+jω) with ε = 0.1 and $\omega_0 = 1$. The white spots are poles and the black spots are zeroes. All 16 poles are shown. Each zero has multiplicity of two, and 12 zeroes are shown and four are located outside the picture, two on the positive ω axis, and two on the negative. The poles of the transfer function are poles on the left half plane and the zeroes of the transfer function are the zeroes, but with multiplicity 1. Black corresponds to a gain of 0.05 or less, white corresponds to a gain of 20 or more.

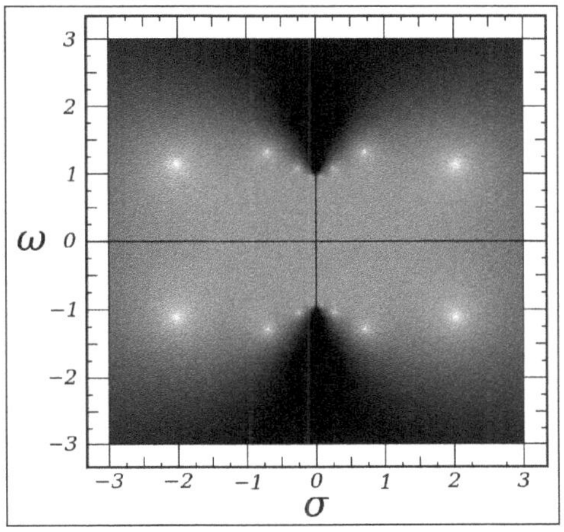

Assuming that the cutoff frequency is equal to unity, the poles (ω_{pm}) of the gain of the Chebyshev filter are the zeroes of the denominator of the gain:

$$1+\varepsilon^2 T_n^2(-1/js_{pm})=0.$$

The poles of gain of the type II Chebyshev filter are the inverse of the poles of the type I filter:

$$\frac{1}{s_{pm}^{\pm}}=\pm\sinh\left(\frac{1}{n}\operatorname{arsinh}\left(\frac{1}{\varepsilon}\right)\right)\sin(\theta_m)$$

$$+j\cosh\left(\frac{1}{n}\operatorname{arsinh}\left(\frac{1}{\varepsilon}\right)\right)\cos(\theta_m)$$

where m = 1, 2,..., n. The zeroes (ω_{zm}) of the type II Chebyshev filter are the zeroes of the numerator of the gain:

$$\varepsilon^2 T_n^2(-1/js_{zm})=0.$$

The zeroes of the type II Chebyshev filter are therefore the inverse of the zeroes of the Chebyshev polynomial.

$$1/s_{zm}=-j\cos\left(\frac{\pi}{2}\frac{2m-1}{n}\right)$$

for m = 1, 2,..., n.

The Transfer Function

The transfer function is given by the poles in the left half plane of the gain function, and has the same zeroes but these zeroes are single rather than double zeroes.

The Group Delay

The gain and the group delay for a fifth-order type II Chebyshev filter with ε=0.1 are plotted in the graph on the left. It can be seen that there are ripples in the gain in the stopband but not in the pass band.

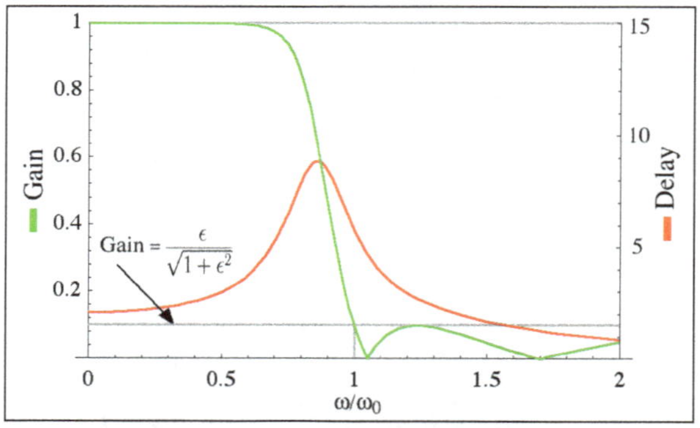

Gain and group delay of a fifth-order type II Chebyshev filter with ε = 0.1.

Implementation

Cauer Topology

A passive LC Chebyshev low-pass filter may be realized using a Cauer topology. The inductor or capacitor values of a nth-order Chebyshev prototype filter may be calculated from the following equations:

$$G_0 = 1$$

$$G_1 = \frac{2A_1}{\gamma}$$

$$G_k = \frac{4A_{k-1}A_k}{B_{k-1}G_{k-1}}, \qquad k = 2,3,4,\ldots,n$$

$$G_{n+1} = \begin{cases} 1 & \text{if } n \text{ odd} \\ \coth^2\left(\frac{\beta}{4}\right) & \text{if } n \text{ even} \end{cases}$$

G1, Gk are the capacitor or inductor element values. fH, the 3 dB frequency is calculated with:

$$f_H = f_0 \cosh\left(\frac{1}{n}\cosh^{-1}\frac{1}{\varepsilon}\right)$$

The coefficients A, γ, β, A_k, and B_k may be calculated from the following equations:

$$\gamma = \sinh\left(\frac{\beta}{2n}\right)$$

$$\beta = \ln\left[\coth\left(\frac{\delta}{17.37}\right)\right]$$

$$A_k = \sin\frac{(2k-1)\pi}{2n}, \qquad k = 1,2,3,\ldots,n$$

$$B_k = \gamma^2 + \sin^2\left(\frac{k\pi}{n}\right), \qquad k = 1,2,3,\ldots,n$$

where δ is the passband ripple in decibels. The number 17.37 is rounded from the exact value $40/\ln(10)$.

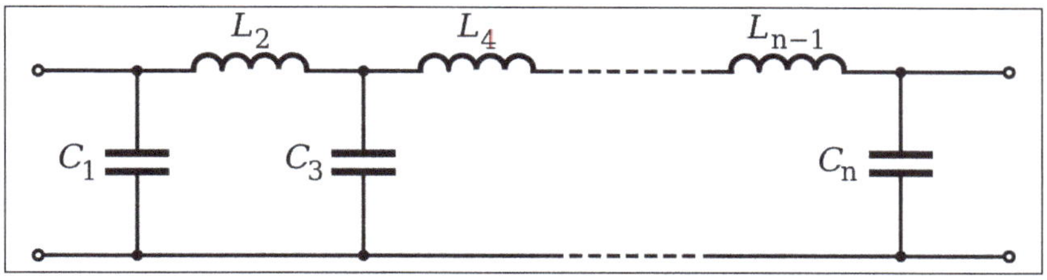

Low-pass filter using Cauer topology.

The calculated G_k values may then be converted into shunt capacitors and series inductors as shown on the right, or they may be converted into series capacitors and shunt inductors. For example,

- $C_{1\ shunt} = G_1$, $L_{2\ series} = G_2$,...

or

- $L_{1\ shunt} = G_1$, $C_{1\ series} = G_2$,...

Note that when G_1 is a shunt capacitor or series inductor, G_0 corresponds to the input resistance or conductance, respectively. The same relationship holds for G_{n+1} and G_n. The resulting circuit is a normalized low-pass filter. Using frequency transformations and impedance scaling, the normalized low-pass filter may be transformed into high-pass, band-pass, and band-stop filters of any desired cutoff frequency or bandwidth.

Digital

As with most analog filters, the Chebyshev may be converted to a digital (discrete-time) recursive form via the bilinear transform. However, as digital filters have a finite bandwidth, the response shape of the transformed Chebyshev is warped. Alternatively, the Matched Z-transform method may be used, which does not warp the response.

Comparison with other Linear Filters

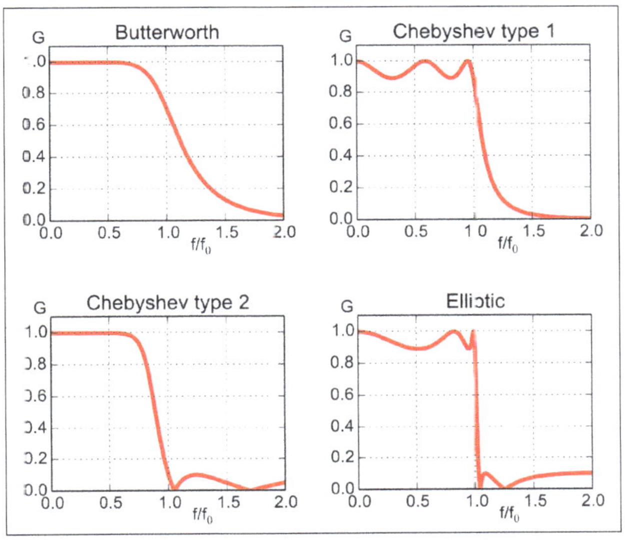

Chebyshev filters are sharper than the Butterworth filter; they are not as sharp as the elliptic one, but they show fewer ripples over the bandwidth.

The Chebyshev filters next to other common filter types obtained with the same number of coefficients (fifth order).

Butterworth Filter

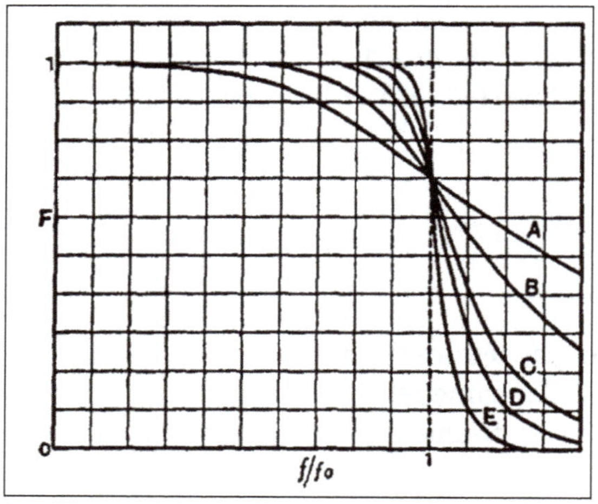

The frequency response plot from Butterworth's 1930 paper.

The Butterworth filter is a type of signal processing filter designed to have a frequency response as flat as possible in the passband. It is also referred to as a maximally flat magnitude filter. It was first described in 1930 by the British engineer and physicist Stephen Butterworth in his paper entitled "On the Theory of Filter Amplifiers".

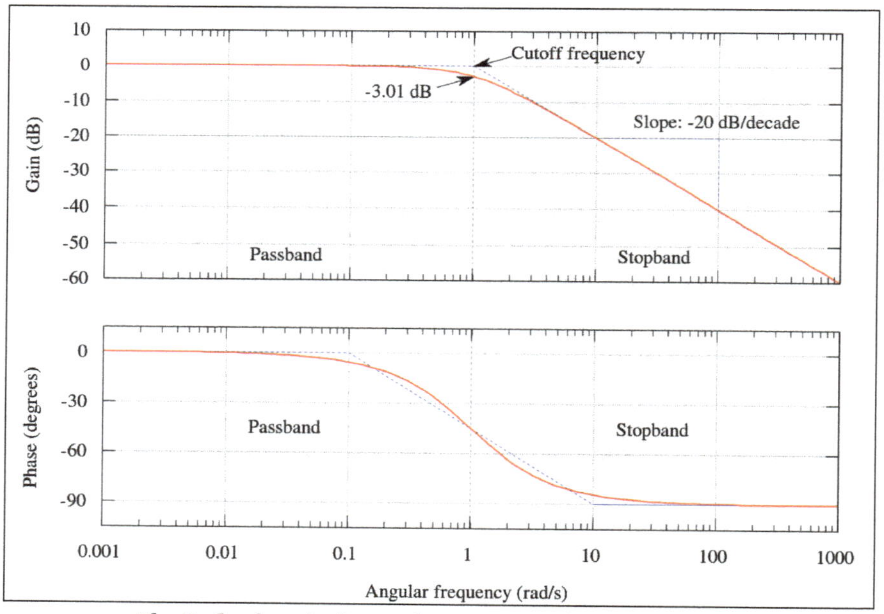

The Bode plot of a first-order Butterworth low-pass filter.

The frequency response of the Butterworth filter is maximally flat (i.e. has no ripples) in the passband and rolls off towards zero in the stopband. When viewed on a logarithmic Bode plot, the response slopes off linearly towards negative infinity. A first-order filter's response rolls off at −6 dB per octave (−20 dB per decade) (all first-order lowpass filters have the same normalized frequency response). A second-order filter decreases at −12 dB per octave, a third-order at −18 dB and so on. Butterworth filters have a monotonically changing magnitude function with ω, unlike other filter types that have non-monotonic ripple in the passband and the stopband.

Compared with a Chebyshev Type I/Type II filter or an elliptic filter, the Butterworth filter has a slower roll-off, and thus will require a higher order to implement a particular stopband specification, but Butterworth filters have a more linear phase response in the pass-band than Chebyshev Type I/Type II and elliptic filters can achieve.

Example:

A transfer function of a third-order low-pass Butterworth filter design shown in the figure on the right looks like this:

$$\frac{V_o(s)}{V_i(s)} = \frac{R}{s^3(L_1C_2L_3) - s^2(L_1C_2R) + s(L_1 + L_3) + R}.$$

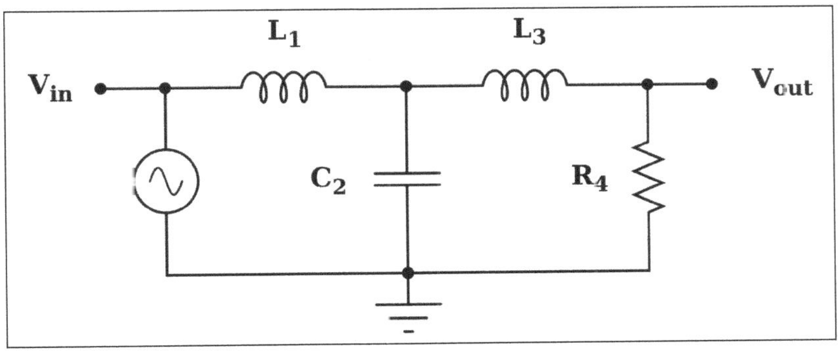

A third-order low-pass filter (Cauer topology). The filter becomes a Butterworth filter with cutoff frequency ω_c=1 when (for example) C_2=4/3 F, R_4=1 Ω, L_1=3/2 H and L_3=1/2 H.

A simple example of a Butterworth filter is the third-order low-pass design shown in the figure on the right, with C_2 = 4/3 F, R_4 = 1 Ω, L_1 = 3/2 H, and L_3 = 1/2 H. Taking the impedance of the capacitors C to be 1/(Cs) and the impedance of the inductors L to be Ls, where s = σ + jω is the complex frequency, the circuit equations yield the transfer function for this device:

$$H(s) = \frac{V_o(s)}{V_i(s)} = \frac{1}{1 + 2s + 2s^2 + s^3}.$$

The magnitude of the frequency response (gain) G(ω) is given by:

$$G(\omega) = |H(j\omega)| = \frac{1}{\sqrt{1+\omega^6}},$$

obtained from,

$$G^2(\omega) = |H(j\omega)|^2 = H(j\omega) \cdot H^*(j\omega) = \frac{1}{1-\omega^6},$$

and the phase is given by:

$$\text{Ö}(\omega) = \arg(H(j\omega)).$$

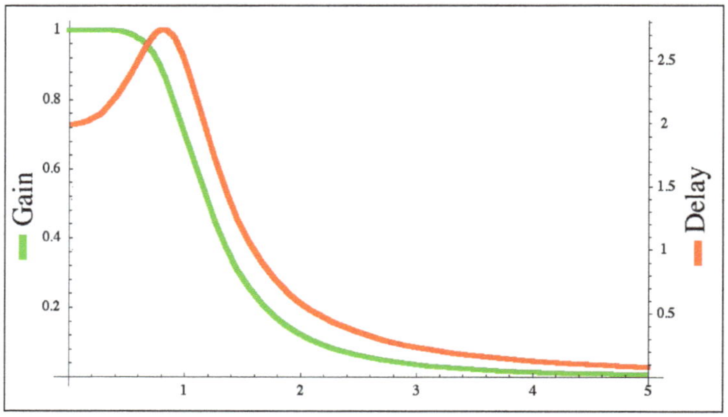

Gain and group delay of the third-order Butterworth filter with ω_c=1.

The group delay is defined as the derivative of the phase with respect to angular frequency and is a measure of the distortion in the signal introduced by phase differences for different frequencies. The gain and the delay for this filter are plotted in the graph on the left. It can be seen that there are no ripples in the gain curve in either the passband or the stop band.

The log of the absolute value of the transfer function *H(s)* is plotted in complex frequency space in the second graph on the right. The function is defined by the three poles in the left half of the complex frequency plane.

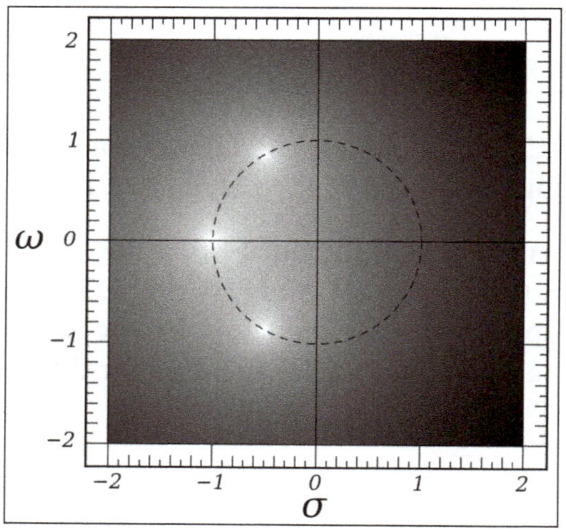

Log density plot of the transfer function H(s) in complex frequency space for the third-order Butterworth filter with ω_c=1. The three poles lie on a circle of unit radius in the left half-plane.

These are arranged on a circle of radius unity, symmetrical about the real s axis. The gain function will have three more poles on the right half plane to complete the circle.

By replacing each inductor with a capacitor and each capacitor with an inductor, a high-pass Butterworth filter is obtained.

A band-pass Butterworth filter is obtained by placing a capacitor in series with each inductor and an inductor in parallel with each capacitor to form resonant circuits. The value of each new component must be selected to resonate with the old component at the frequency of interest.

A band-stop Butterworth filter is obtained by placing a capacitor in parallel with each inductor and an inductor in series with each capacitor to form resonant circuits. The value of each new component must be selected to resonate with the old component at the frequency to be rejected.

Transfer Function

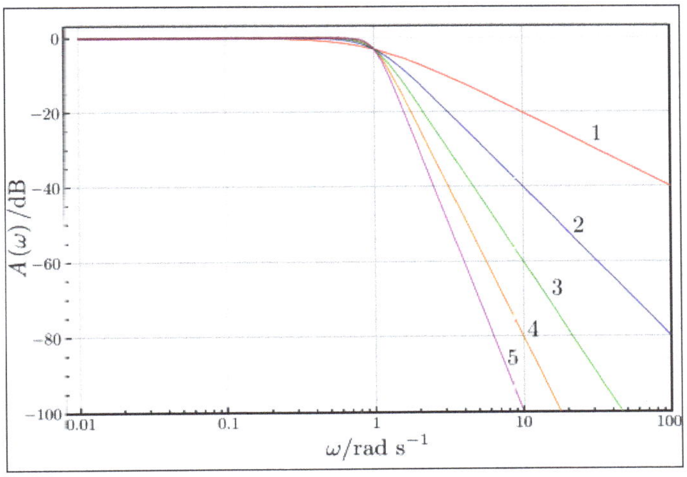

Plot of the gain of Butterworth low-pass filters of orders 1 through 5, with cutoff frequency $\omega_o = 1$. Note that the slope is 20n dB/decade where n is the filter order.

Like all filters, the typical prototype is the low-pass filter, which can be modified into a high-pass filter, or placed in series with others to form band-pass and band-stop filters, and higher order versions of these.

The gain $G(\omega)$ of an n-order Butterworth low-pass filter is given in terms of the transfer function $H(s)$ as

$$G^2(\omega) = |H(j\omega)|^2 = \frac{G_o^{\ 2}}{1+(\frac{j\omega}{j\omega_c})^{2n}}$$

where

- n = order of filter.
- ω_c = cutoff frequency (approximately the -3dB frequency).
- G_o is the DC gain (gain at zero frequency).

It can be seen that as n approaches infinity, the gain becomes a rectangle function and frequencies below ω_c will be passed with gain G_o, while frequencies above ω_c will be suppressed. For smaller values of n, the cutoff will be less sharp.

We wish to determine the transfer function H(s) where $s=\sigma+j\omega$ (from Laplace transform). Because $|H(s)|^2=H(s)\overline{H(s)}$ and, as a general property of Laplace transforms at $s=j\omega$,, $H(-j\omega)=\overline{H(j\omega)}$, if we select H(s) such that:

$$H(s)H(-s)=\frac{G_0^{\,2}}{1+\left(\frac{-s^2}{\omega_c^2}\right)^n},$$

then, with $s=j\omega$,, we have the frequency response of the Butterworth filter.

The n poles of this expression occur on a circle of radius ω_c at equally-spaced points, and symmetric around the negative real axis. For stability, the transfer function, H(s), is therefore chosen such that it contains only the poles in the negative real half-plane of s. The k-th pole is specified by,

$$-\frac{s_k^2}{\omega_c^2}=(-1)^{\frac{1}{n}}=e^{\frac{j(2k-1)\pi}{n}} \qquad k=1,2,3,\dots,n$$

and hence;

$$s_k=\omega_c e^{\frac{j(2k+n-1)\pi}{2n}} \qquad k=1,2,3,\dots,n.$$

The transfer(or system) function may be written in terms of these poles as,

$$H(s)=\frac{G_0}{\prod_{k=1}^{n}(s-s_k)/\omega_c}$$

Where $\prod$ is the product of a sequence operator. The denominator is a Butterworth polynomial in s.

Normalized Butterworth Polynomials

The Butterworth polynomials may be written in complex form as above, but are usually written with real coefficients by multiplying pole pairs that are complex conjugates, such as s_1 and s_n. The polynomials are normalized by setting $\omega_c=1$. The normalized Butterworth polynomials then have the general form

$$B_n(s)=\prod_{k=1}^{\frac{n}{2}}\left[s^2-2s\cos\left(\frac{2k+n-1}{2n}\pi\right)+1\right] \qquad n=even$$

$$B_n(s)=(s+1)\prod_{k=1}^{\frac{n-1}{2}}\left[s^2-2s\cos\left(\frac{2k+n-1}{2n}\pi\right)+1\right] \qquad n=odd.$$

To four decimal places, they are

n	Factors of Polynomial
1	$(s+1)$

2	$(s^2+1.4142s+1)$
3	$(s+1)(s^2+s+1)$
4	$(s^2+0.7654s+1)(s^2+1.8478s+1)$
5	$(s+1)(s^2+0.6180s+1)(s^2+1.6180s+1)$
6	$(s^2+0.5176s+1)(s^2+1.4142s+1)(s^2+1.9319s+1)$
7	$(s+1)(s^2+0.4450s+1)(s^2+1.2470s+1)(s^2+1.8019s+1)$
8	$(s^2+0.3902s+1)(s^2+1.1111s+1)(s^2+1.6629s+1)(s^2+1.9616s+1)$
9	$(s+1)(s^2+0.3473s+1)(s^2+s+1)(s^2+1.5321s+1)(s^2+1.879s+1)$
10	$(s^2+0.3129s+1)(s^2+0.9080s+1)(s^2+1.4142s+1)(s^2+1.7820s+1)(s^2+1.9754s+1)$

The normalized Butterworth polynomials can be used to determine the transfer function for any low-pass filter cut-off frequency ω_c, as follows,

$H(s)=\frac{G_0}{B_n(a)}$, where $a=\frac{s}{\omega_c}$.

Transformation to other bandforms are also possible,

Maximal Flatness

Assuming $\omega_c=1$ and $G_0=1$, the derivative of the gain with respect to frequency can be shown to be,

$$\frac{dG}{d\omega}=-nG^3\omega^{2n-1}$$

which is monotonically decreasing for all ω since the gain G is always positive. The gain function of the Butterworth filter therefore has no ripple. The series expansion of the gain is given by

$$G(\omega)=1-\frac{1}{2}\omega^{2n}+\frac{3}{8}\omega^{4n}+\dots$$

In other words, all derivatives of the gain up to but not including the 2n-th derivative are zero at $\omega=0$, resulting in "maximal flatness". If the requirement to be monotonic is limited to the passband only and ripples are allowed in the stopband, then it is possible to design a filter of the same order, such as the inverse Chebyshev filter, that is flatter in the passband than the "maximally flat" Butterworth.

High-frequency Roll-off

Again assuming $\omega_c=1$, the slope of the log of the gain for large ω is,

$$\lim_{\omega\to\infty}\frac{d\log(G)}{d\log(\omega)}=-n.$$

In decibels, the high-frequency roll-off is therefore 20n dB/decade, or 6n dB/octave (the factor of 20 is used because the power is proportional to the square of the voltage gain).

Filter Implementation and Design

There are several different filter topologies available to implement a linear analogue filter. The most often used topology for a passive realisation is Cauer topology and the most often used topology for an active realisation is Sallen–Key topology.

Cauer Topology

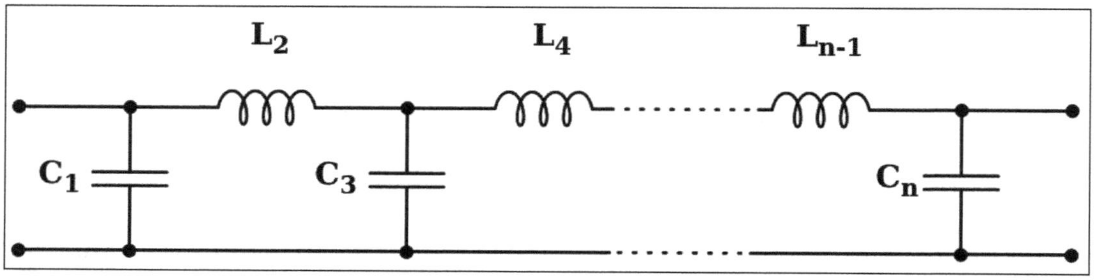

Butterworth filter using Cauer topology.

The Cauer topology uses passive components (shunt capacitors and series inductors) to implement a linear analog filter. The Butterworth filter having a given transfer function can be realised using a Cauer 1-form. The k-th element is given by:

$$C_k = 2\sin\left[\frac{(2k-1)}{2n}\pi\right] \qquad k = \text{odd}$$

$$L_k = 2\sin\left[\frac{(2k-1)}{2n}\pi\right] \qquad k = \text{even}.$$

The filter may start with a series inductor if desired, in which case the L_k are k odd and the C_k are k even. These formulae may usefully be combined by making both L_k and C_k equal to g_k. That is, g_k is the immittance divided by s.,

$$g_k = 2\sin\left[\frac{(2k-1)}{2n}\pi\right] \qquad k = 1,2,3,\ldots,n.$$

These formulae apply to a doubly terminated filter (that is, the source and load impedance are both equal to unity) with $\omega_c = 1$. This prototype filter can be scaled for other values of impedance and frequency. For a singly terminated filter (that is, one driven by an ideal voltage or current source) the element values are given by:

$$g_j = \frac{a_j a_{j-1}}{c_{j-1} g_{j-1}} \qquad j = 2,3,\ldots,n$$

where,

$$g_1 = a_1$$

and

$$a_j = \sin\frac{\pi}{2}\left[\frac{(2j-1)}{n}\right] \qquad j = 1,2,3,\ldots,n$$

$$c_j = \cos^2\left[\frac{\pi j}{2n}\right] \qquad j = 1,2,3,\ldots,n.$$

Voltage driven filters must start with a series element and current driven filters must start with a shunt element. These forms are useful in the design of diplexers and multiplexers.

Sallen–Key Topology

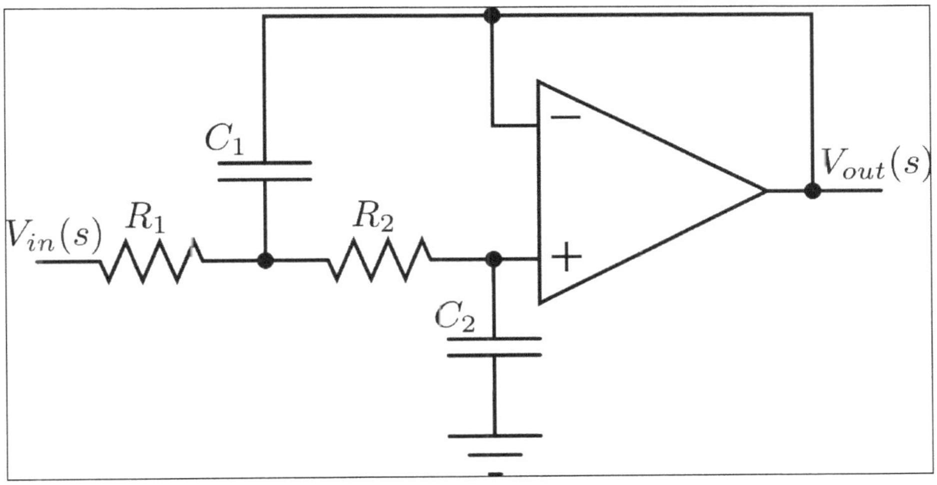

Sallen–Key topology.

The Sallen–Key topology uses active and passive components (noninverting buffers, usually op amps, resistors, and capacitors) to implement a linear analog filter. Each Sallen–Key stage implements a conjugate pair of poles; the overall filter is implemented by cascading all stages in series. If there is a real pole (in the case where n is odd), this must be implemented separately, usually as an RC circuit, and cascaded with the active stages.

For the second-order Sallen–Key circuit shown to the right the transfer function is given by:

$$H(s) = \frac{V_{out}(s)}{V_{in}(s)} = \frac{1}{1 + C_2(R_1 + R_2)s + C_1C_2R_1R_2s^2}.$$

We wish the denominator to be one of the quadratic terms in a Butterworth polynomial. Assuming that $\omega_c = 1$, this will mean that,

$$C_1C_2R_1R_2 = 1$$

and

$$C_2(R_1 + R_2) = -2\cos\left(\frac{2k+n-1}{2n}\pi\right).$$

This leaves two undefined component values that may be chosen at will.

Digital Implementation

Digital implementations of Butterworth and other filters are often based on the bilinear transform method or the matched Z-transform method, two different methods to discretize an analog filter design. In the case of all-pole filters such as the Butterworth, the matched Z-transform method is equivalent to the impulse invariance method. For higher orders, digital filters are sensitive to quantization errors, so they are often calculated as cascaded biquad sections, plus one first-order or third-order section for odd orders.

Comparison with other Linear Filters

Properties of the Butterworth filter are:

- Monotonic amplitude response in both passband and stopband.
- Quick roll-off around the cutoff frequency, which improves with increasing order.
- Considerable overshoot and ringing in step response, which worsens with increasing order.
- Slightly non-linear phase response.
- Group delay largely frequency-dependent.

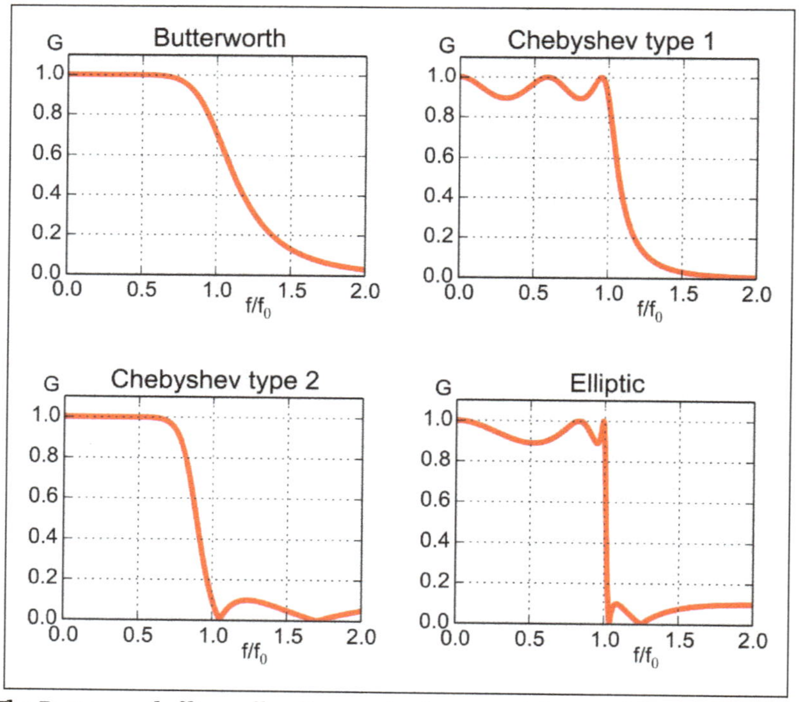

The Butterworth filter rolls off more slowly around the cutoff frequency than the Chebyshev filter or the Elliptic filter, but without ripple.

Here is an image showing the gain of a discrete-time Butterworth filter next to other common filter types. All of these filters are fifth-order.

References

- Miroslav D. Lutovac, Dejan V. Tošić, Brian Lawrence Evans, Filter Design for Signal Processing Using MATLAB and Mathematica, Miroslav Lutovac, 2001 ISBN 0201361302
- Andrews, Keith; posting as ssltech (January 11, 2010). "Re: Running the board for a show this big?". Recording, Engineering & Production. Prosoundweb. Archived from the original on 15 July 2011. Retrieved 9 March 2010
- L. D. Paarmann, Design and Analysis of Analog Filters: A Signal Processing Perspective, Springer, 2001 ISBN 0792373731
- Paul M. Mather (2004). Computer processing of remotely sensed images: an introduction (3rd ed.). John Wiley and Sons. P. 181. ISBN 978-0-470-84919-4
- Michael Goesele (2004). New Acquisition Techniques for Real Objects and Light Sources in Computer Graphics. Books on Demand. P. 34. ISBN 978-3-8334-1489-3
- Band-stop-filter, filter: electronics-tutorials.ws, Retrieved 8 January, 2019
- Belle A. Shenoi (2006). Introduction to digital signal processing and filter design. John Wiley and Sons. P. 120. ISBN 978-0-471-46482-2

Permissions

All chapters in this book are published with permission under the Creative Commons Attribution Share Alike License or equivalent. Every chapter published in this book has been scrutinized by our experts. Their significance has been extensively debated. The topics covered herein carry significant information for a comprehensive understanding. They may even be implemented as practical applications or may be referred to as a beginning point for further studies.

We would like to thank the editorial team for lending their expertise to make the book truly unique. They have played a crucial role in the development of this book. Without their invaluable contributions this book wouldn't have been possible. They have made vital efforts to compile up to date information on the varied aspects of this subject to make this book a valuable addition to the collection of many professionals and students.

This book was conceptualized with the vision of imparting up-to-date and integrated information in this field. To ensure the same, a matchless editorial board was set up. Every individual on the board went through rigorous rounds of assessment to prove their worth. After which they invested a large part of their time researching and compiling the most relevant data for our readers.

The editorial board has been involved in producing this book since its inception. They have spent rigorous hours researching and exploring the diverse topics which have resulted in the successful publishing of this book. They have passed on their knowledge of decades through this book. To expedite this challenging task, the publisher supported the team at every step. A small team of assistant editors was also appointed to further simplify the editing procedure and attain best results for the readers.

Apart from the editorial board, the designing team has also invested a significant amount of their time in understanding the subject and creating the most relevant covers. They scrutinized every image to scout for the most suitable representation of the subject and create an appropriate cover for the book.

The publishing team has been an ardent support to the editorial, designing and production team. Their endless efforts to recruit the best for this project, has resulted in the accomplishment of this book. They are a veteran in the field of academics and their pool of knowledge is as vast as their experience in printing. Their expertise and guidance has proved useful at every step. Their uncompromising quality standards have made this book an exceptional effort. Their encouragement from time to time has been an inspiration for everyone.

The publisher and the editorial board hope that this book will prove to be a valuable piece of knowledge for students, practitioners and scholars across the globe.

Index

CPSIA information can be obtained
at www.ICGtesting.com
Printed in the USA
LVHW060902030122
707669LV00002B/305

9 781647 283315